Workability and Quality Control of Concrete

Other books on concrete available from E & FN Spon

Reinforced Concrete Design to BS 8110 – Simply explained
A.H. Allen

Rheology of Fresh Cement and Concrete
Edited by P.F.G. Banfill

Protection of Concrete
Edited by R.K. Dhir and J.W. Green

Reinforced Concrete Design Theory and Examples, Second Edition
T.J. Machinley and B.S. Choo

Testing during Concrete Construction
Edited by H.W. Reinhardt

Reinforced Concrete Designer's Handbook, Tenth Edition
C.E. Reynolds and J.C. Steedman

Supervision of Concrete Construction, Volumes 1 and 2
J.G. Richardson

Chemical Admixtures for Concrete, Second Edition
M.R. Rixom and N.P. Mailvaganam

Corrosion of Steel in Concrete
Edited by P. Schiessl

Admixtures for Concrete – Improvement of Properties
Edited by E. Vazquez

Properties of Fresh Concrete
Edited by H.-J. Wierig

Workability and Quality Control of Concrete

G.H. TATTERSALL
M.Sc., Ph.D., C.Phys., F.Inst.P., F.I. Ceram.

Honorary Reader
Department of Civil and Structural Engineering
University of Sheffield

E & FN SPON
An Imprint of Chapman & Hall
London · New York · Tokyo · Melbourne · Madras

Published by E & FN Spon, an imprint of Chapman & Hall, 2–6 Boundary Row, London SE1 8HN

Chapman & Hall, 2–6 Boundary Row, London SE1 8HN, UK

Van Nostrand Reinhold Inc., 115 5th Avenue, New York NY10003, USA

Chapman & Hall Japan, Thomson Publishing Japan, Hirakawacho Nemoto Building, 7F, 1-7-11 Hirakawa-cho, Chiyoda-ku, Tokyo 102, Japan

Chapman & Hall Australia, Thomas Nelson Australia, 102 Dodds Street, South Melbourne, Victoria 3205, Australia

Chapman & Hall India, R. Seshadri, 32 Second Main Road, CIT East, Madras 600 035, India

First edition 1991

© 1991 G.H. Tattersall

Typeset in 10/12pt Palatino by Excel Typesetter Company
Printed in Great Britain by T.J. Press (Padstow) Ltd, Padstow, Cornwall

ISBN 0 419 14860 4   0 442 31245 8 (USA)

A catalogue record for this book is available from the British Library

Library of Congress Cataloging-in-Publication data
Tattersall, G.H.
     Workability and quality control of concrete/G.H. Tattersall.
          p.   cm.
     Includes bibliographical references and index.
     ISBN 0-442-31245-8
     1. Concrete—Testing.   2. Concrete—Quality control.   I. Title.
TA440.T27   1991
620.1'360287—dc20   90-26033
                                                                    CIP

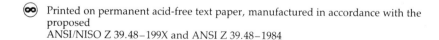

Printed on permanent acid-free text paper, manufactured in accordance with the proposed
ANSI/NISO Z 39.48–199X and ANSI Z 39.48–1984

Contents

Preface

This book is a logical follow-up of the monograph *The Workability of Concrete* which was published in 1976, and which was inspired by a suggestion from the then Cement and Concrete Association, as a result of lectures I gave for some twenty years on the Advanced Concrete Technology Course held at the Association's Training Centre at Fulmer Grange. Since 1976 a great deal of work has been done in studies of workability measurement and its applications, and it has been demonstrated beyond doubt that the flow properties of fresh concrete can be satisfactorily represented by the Bingham model. This finding has profound implications for the industry and my main purpose is to discuss and elucidate those implications.

However, workability and the factors that affect it are considered also in terms of the tests that are already familiar, although the disadvantages of those methods are argued strongly. Like my lectures, and like the earlier book, this one is intended for practising engineers and concrete technologists, and I hope it will be seen as proposing procedures that are eminently practical, although they may at first seem to be a little more complicated than those at present in use. In fact, very little extra trouble is required and any that is should be balanced against the benefits to be obtained in terms of a better understanding of fresh concrete, and in increased hope of avoiding the difficulties and disputes that are expensive in time and money.

There is very little mathematics in the book. I have not provided proofs of any equations used but have tried to present reasonable arguments that I hope will appeal to those who are engaged in getting the job done, rather than in carrying out interesting investigations in the absence of pressure. I have also, on occasion, simplified the physics arguments but without sacrificing anything that is of importance in practical application. Any reader who feels dissatisfied about lack of rigour in any section is invited to refer to the original papers where full arguments are given.

G.H. Tattersall

Acknowledgments

It is impossible to list all those who have contributed to the work considered in this book. Those whose papers are considered specifically are of course named in the text but I should like to thank all those scores of people who have discussed and co-operated with me and perhaps I may mention particularly those who have been members of my own research team.

The work would have been impossible without the ready and willing help of those in industry who have provided facilities and materials free of charge. They have included consultants, contractors, ready-mixed concrete suppliers, and manufacturers of cement, admixtures, cement replacements and equipment. I hope they in particular will feel that their contributions have not been wasted.

Finally I should like to thank Mr Nick Clarke my editor, Dr Peter Wainwright of the University of Leeds, and my son Mr David Tattersall of Ove Arup & Partners for reading through the typescript and making helpful comments. They certainly helped to improve the text but I did not accept everything they said and I take full responsibility for the result.

G.H.T.

Note

I have used some statistical terms with which some readers may not be familiar. An understanding of them is not essential to an understanding of the main thesis of the book but I have provided explanations in a glossary. I have also provided a glossary of some rheological terms.

Anyone wishing to use any of the standard specifications or tests discussed is advised to refer to the appropriate Standard for operational details. Extracts from British Standards are reproduced with the kind permission of the British Standards Institution. Complete copies can be obtained by post from BSI Sales, Linford Wood, Milton Keynes MK14 6LE (telex 825777 BSIMK G).

1 The importance of workability

1.1 THE MEANING OF WORKABILITY

The word **workability** is a term that refers to properties of fresh concrete, that is, of the concrete before it has set and hardened, and it is legitimate to ask why any attention should be given to these properties at all. The performance of concrete will in practice be assessed in terms of whether the hardened material performs in the way intended and continues to do so: it will be judged in terms of shape and finish, strength, deflection, dimensional changes, permeability and durability. So why should the properties of the fresh concrete be considered to be important, and why should they be the concern of the practising engineer?

The answer to the first of these questions lies in the fact that the properties of any finished material are affected by the properties at an earlier stage and by the processes applied to it, while the answer to the second one is that all, or a major part of, the processing of concrete is actually carried out on site. The first stage is, of course, the making of a homogeneous mix and then, assuming this has been done properly, the material is subjected to other processes as follows.

(a) Transport

The concrete must be capable of being transported and placed by any one or more of a variety of methods including barrow, dumper truck, skip, mixer truck, conveyor belt, pumping and tremie.

(b) Flow

It must be capable of flowing into all corners of a mould or formwork to fill it completely, and this process may be rendered more difficult by the presence of congested reinforcement or awkward sections (see Figure 1.1).

Figure 1.1 Congested reinforcement (*Photo Courtesey of Ove Arup Partnership*).

(c) *Compaction*

It must be capable of being thoroughly compacted to expel air to achieve the maximum potential strength and durability of the hardened concrete. The presence of 5% residual air reduces strength by about 30%, and 10% air results in a loss of strength of more than half. Compaction is sometimes carried out manually but it is normal practice to use vibrators or, for precast products, such processes as hydraulic pressing or extrusion (see Figure 1.2).

(d) Finishing

The concrete must be capable of giving a good finish direct from the formwork, without honeycombing or an excessive number of blowholes or other surface defects. If there is a free surface, it must also be capable of giving a good finish in response to an operation such as floating or trowelling.

A workable concrete is one that satisfies these requirements without difficulty and, in general, the more workable it is, that is, the higher its workability, the more easily it can be placed, compacted and finished. Workability can be increased by simply increasing the water content of the mix but, if that method is used, a point will be reached at which segregation and/or bleeding become unacceptable so that the concrete is no longer homogeneous and, before that, the water/cement ratio may have reached a level such that the hardened concrete will

Figure 1.2 Compaction with poker vibrations (*Photo Courtesy of Ove Arup Partnership*).

not attain the required strength. If, on the other hand, workability is increased by increasing both the water content and the cement content to avoid the problems of a heightened water/cement ratio, the cost of the material will be greater and there will be more severe problems with heat evolution, shrinkage, and creep.

Thus there are conflicting factors and the technology (and art) of mix design and specification lies in reconciling them. If there are no further restrictions, such as a required minimum cement content, the most economical mix is the one for which the workability is as low as possible, while still permitting the objectives listed above to be achieved. Clearly the minimum workability necessary will depend very much on the nature of the job to be done; concrete of extremely low workability may be suitable for road construction with paving trains that utilize vibration to aid in compaction, but a much higher workability will be needed for the tremieing of diaphragm walls.

It follows that there are two important practical questions that must be answered:

(a)   What is the minimum workability that is necessary for the particular job in hand?

(b)   What specification will produce a mix that possesses at least that minimum workability and also satisfies other requirements, such as those concerning the strength and durability of the hardened concrete?

These two questions lie at the basis of all methods of mix design. To answer them properly it is necessary first:

(a)   to define unambiguously what is meant by workability;

(b)   to define a scale of measurement;

(c)   to provide a method of measurement, based on the definitions;

(d)   to provide, in terms of measurements so obtained, quantitative information to describe completely concretes suitable for particular jobs;

(e)   to indicate how desired levels of workability may be achieved, by providing information on the effects on workability of the properties and proportions of mix constituents.

These are the bare minimum requirements to provide a fully satisfactory basis for mix design and quality control, but there is another highly desirable one related to control:

(f)   to be able to indicate, if workability is not as specified, why it is not, that is, what unintended change in the properties or proportions of mix constituents is responsible for the failure to meet the specification.

These six requirements should be borne in mind; the ways in which they have been dealt with, or ignored, will be considered in detail.

1.2 DEFINITION OF WORKABILITY

In some areas of concrete work there may be a conscious decision not to attempt at all the difficult task of defining what is seen as a complex property, and workability is referred to simply in terms such as 'high', 'medium', or 'low', or mixes are described as being 'semi-dry' or 'plastic' and so on. These descriptions may be of some use in very restricted circumstances and for communication between very restricted numbers of people, but it is evident that they are of little use for dealing with the problem in general, and there is no guarantee that any particular one of these terms means the same thing to different users. They are at best extremely vague.

Even worse is the **unconscious** decision, or tacit assumption, that no definition is necessary, and it is readily apparent that such is the inadequate basis for many of the plethora of proposals for empirical tests by methods that lack either any theoretical basis or any connection with practice. Thinking seems to have been on the very superficial level that if the new test seems to indicate a change in workability in the same direction (i.e. higher or lower) as that observed subjectively, it must have some merit.

The word 'workability' obviously has the meaning 'the ability to be worked', but to define it in such a way is to present a tautology that is of no practical value. Consideration of this meaning has however led some workers to attempt a definition of workability that includes a consideration not only of the concrete itself but also of the job in which it is to be used. For example, Baron and Lesage[1] suggested a definition based on assessment of the porosity of concrete that had been laid in a particular way. Those who argue that the definition should be of this form, so that the workability of a given concrete should be reckoned as different if it were to be used in a different way, might well be asked why they do not apply the same principle to other physical properties. For example, does anybody suggest that the value of the Young's modulus of a material should depend on whether the material is used for a simply supported beam or for a cantilever? Attempts at measuring physical properties should always be aimed at the expression of results in quantitative terms that are independent not only of the way in which the material is to be used, but also independent of the nature of the apparatus used to measure them.

Sometimes, proposed methods for workability measurement have been intended to imitate practical conditions, but in a more or less closely defined way. The VB consistometer, to be considered in detail later, clearly represents an attempt to take account of the use of vibration for placing and compacting. Angles[2] made a very specific statement in support of this type of approach when he said, 'Surely it would not be contested that the most effective measuring device would simulate as far as possible the operation of placing concrete in typical conditions.' In fact, this statement can not only be contested, it can, as will appear later, be shown to be wrong.

The only satisfactory way is to attempt to define workability in the way used for any other physical property, and that means an attempt to satisfy certain criteria as follows:

(a)   Workability is a property of the concrete alone.
(b)   Workability will be expressed quantitatively in terms of one or more physical constants which may be called $W_1, W_2 \ldots W_n$. This set may be written in shorthand form as W_i. There is no a priori way of knowing how many constants there will be in the set. It may be pointed out that all the British Standards, and all other standard tests, involve the underlying assumption that only one constant is needed, i.e. that workability can be expressed quantitatively as a single figure, such as the slump, or the vebe time.
(c)   All the constants in the set W_i should be expressible in terms of the fundamental units of mass, length, and time, or of units derived from them, such as stress and shear rate. In other words, the results should be independent of the apparatus used to obtain them.
(d)   The constants must, together, provide a sufficiently complete description: that is, if two concretes have the same numerical values of all the necessary constants, then those two concretes must behave in the same way in any given set of practical circumstances.

In practice, and for practical industrial purposes, it may be necessary, and acceptable, to settle for something less than perfect adherence to these criteria. If the number of constants needed for a proper description of workability were large, the problems in practical applications would probably be intractable, so it may be permissible to use approximations if, by so doing, the number could be reduced, but only provided that criterion (d) is not violated. Similarly, although it is highly desirable that results be expressible in fundamental units, the somewhat lesser requirement that they be expressible in terms that are understandable in relation to fundamental properties would suffice in practice. This would mean that results would not be independent of the design details of the apparatus used to measure them, but the apparatus can be standardised so that results from separate set-ups agree.

There can, however, be no compromise concerning the fourth criterion, because any test that is capable of classifying as identical in properties two concretes that may subsequently be found to behave differently on the job, is inadequate even as a simple pass/fail test. A testing deficiency of this sort can be responsible for the sort of expensive mistakes that can happen too frequently in the construction industry.

1.3 TERMINOLOGY

Progress in any sphere is likely to be inhibited if the words used mean different things to different people or if, on the Humpty Dumpty principle, an individual author can choose whatever meaning he happens to like at the time. There is no doubt that, in the field of workability measurement, much confusion has been caused by careless and inaccurate use of terminology and by the formulation of arbitrary definitions that are in conflict with the equally arbitrary definitions of others, or with established usage.

On the one hand, terms like workability itself, which was originally a general descriptive word, and consistency, which is defined only by the complete flow/force curve of a material, have had their meanings restricted in undesirable ways, while, on the other hand, words like viscosity and mobility, which properly do have restricted and rigorously defined meanings, have been used to describe the results from quite empirical tests. This is clearly a very unsatisfactory state of affairs and attention should be given to standardization of terminology.

Unless and until it becomes possible to define workability with rigorous adherence to the criteria listed earlier, the term should be retained for the most general use without quantification. Thus it is valid to refer to high, medium, or low workability or, in particular circumstances, to make statements of the type 'this concrete is more workable than that one', but not to attempt to put numbers to the description. It may also be useful, on occasions, to use words like flowability, compactability, stability, and so on, to describe the constituent properties, but again, without any attempt to quantify. These words are clumsy but they all have the merit of ending in '. . . ability', and their meanings are reasonably self-evident.

Results from empirical tests should be quoted quantitatively but with reference to the test. Such tests do not measure workability and it is misleading to quote results as if they do. Thus the slump test measures the slump of a concrete cone that has been cast and released in a particular way; that is all that it measures, so the result should be stated simply as a slump value and referred to as such. Similar considerations apply to all the other standard tests.

So far, two classes of terms have been considered: the general or qualitative, and the specific quantitative, related to the results of empirical tests. There is a third and very important class consisting of those terms that have been defined rigorously in terms of fundamental quantities, and these should not be used in any other way. Some of them will be discussed later; they include viscosity, fluidity, mobility, and yield value[3].

This proposed scheme of terminology[4], which extends suggestions made by Newman[5], Ritchie[6], and Hughes[7], may be summarized as in Table 1.1.

Glossaries of rheological and statistical terms are provided at the end of the book.

1.4 SUBJECTIVE ASSESSMENT OF WORKABILITY

Assessment of workability by subjective judgement of appearance and handling qualities is, of course, a method as old as concrete itself, but it is still in common use today. The batcherman in a ready-mixed concrete plant may observe the behaviour of the concrete in the mixer, or at discharge, in an attempt to ensure that the product is satisfactory, although it may be remarked that in some cases the field of view available to him, or its level of illumination, is quite inadequate. There are also engineers who would back their own personal judgement against the results of any test they have yet met.

Even when objective figures are quoted, it may be considered necessary to supplement the quantitative information with a qualitative

Table 1.1 Scheme of nomenclature for workability

Class I Qualitative	
Workability	
Flowability	To be used only in a general descriptive
Compactability	way without any attempt to quantify
Stability	
Finishability	
Pumpability	
etc.	
Class II Quantitative empirical	
Slump	To be used as a simple quantitative
Compacting factor	statement of behaviour in a particular
Vebe time	set of circumstances
Flow table spread	
etc.	
Class III Quantitative fundamental	
Viscosity	To be used strictly in conformity with
Mobility	the definitions in BS 5168:1975
Fluidity	*Glossary of rheological terms.*
Yield value	
etc.	

description. For example, *Road Note 4*[8], which for many years was the best-known account in the UK of a method of mix design, gave figures for slump and compacting factor of concretes suitable for various types of job, but also used the terms high, medium, low, and very low workability, and Erntroy and Shacklock[9], in their work on high-strength mixes, found it necessary to introduce a further category of extremely low workability. Other methods of mix design, such as those proposed by McIntosh[10], Owens[11], Murdock and Blackledge[12], Hughes[7] and Alexanderson[13] all give a qualitative description of the workabilities required for particular types of job, and quote corresponding values of one or more of slump, compacting factor, and Vebe time.

There is no doubt that subjective assessment of workability, as of many other physical properties, can be a useful guide when carried out by a very experienced man, but it has severe limitations. It cannot readily be quantified, so information is not easy to pass on, and also it must be realised that the description 'experienced' does not imply just a wide general experience but a fairly specific experience of particular mix components and proportions. In *Design of Normal concrete mixes*[14], which is intended as a replacement for *Road Note 4*, there is no attempt to describe the workability for various jobs. Instead, it is stated: 'It is not considered practical for this publication to define the workability required for various types of construction or placing conditions since this is affected by many factors.'*

1.5 EMPIRICAL METHODS OF MEASUREMENT

The limitations of subjective assessment soon made it obvious that more objective methods were needed, and the importance attached to the matter is indicated by the fact that over one hundred different methods have been proposed for measurement of workability. They may be roughly classified as flow tests, remoulding tests, deforming tests, compacting tests, penetration and pull-out tests, drop tests, and mixer tests, with others that do not fit conveniently into one of these categories. They have been listed elsewhere and have been reviewed recently by Wierig[16]; there is nothing to be gained by a further detailed consideration of them here.

In general, these tests lack any sound theoretical basis and they often involve the formulation of quite arbitrary definitions of workability that have been criticised earlier. Another curious feature is that some authors have proposed a new test on the grounds that the established slump test (to be discussed later) is inadequate, and have then sought to justify their new test by showing that it gives results that correlate with those of the slump test. There have been several

---

* See Appendix.

cases in which an author has put forward what he must have thought was a new method, in ignorance of the fact that it was in essentials similar to one that had been suggested one or two decades earlier.

Most of these empirical tests have been used only by the inventor; the ones that have survived and become more widely known are those that have been selected for incorporation in one or more national standards. By far the best known of these is the slump test, which, with the other important ones, will be considered in the next chapter.

1.6 REFERENCES

1. Baron, J. and Lesage, R. (1973) Proposition pour une définition de la maniabilité, in *Fresh Concrete: Important Properties and their Measurement, Proceedings of a RILEM Seminar 22–24 March 1973, Leeds*, **1**, 1,1-1–1,1-27.
2. Angles, J.G. (1975) Measuring workability, *Concrete*, **8**, 25.
3. British Standards Institution (1975). British Standard BS 5168: *Glossary of Rheological Terms*.
4. Tattersall, G.H. (1973) The principles of measurement of workability and a proposed simple two-point test, in *Fresh Concrete: Important Properties and their Measurement, Proceedings of a RILEM Seminar 22–24 March 1973*, **1**, 2,2-1–2,2-33.
5. Newman, K. (1965) Properties of concrete, *Structural Concrete*, **2**(11), 451–82.
6. Ritchie, A.G.B. (1968) The rheology of fresh concrete, *Proceedings of the American Society of Civil Engineers*, **94**, (COl), 55–74.
7. Hughes, B.P. (1968) The rational design of high quality mixes, *Concrete*, **2**(5), 212–22.
8. Road Research Laboratory (1950) *Design of Concrete Mixes, Road Note No.4*, London, HMSO.
9. Erntroy, H.C. and Shacklock, B.W. (1955) Design of high strength concrete mixes. *Proceedings of a Symposium on Mix Design and Quality Control of Concrete*, London, May 1954, 55–73. London: Cement & Concrete Association.
10. McIntosh, J.D. (1964) *Concrete Mix Design*. London: Cement & Concrete Association.
11. Owens, P.L. (1973) *Basic Mix Method. Selection of Proportions for Medium Strength Concretes*. London: Cement and Concrete Association.
12. Murdock, L.J. and Blackledge, G.F. (1968) *Concrete Materials and Practice*, 4th edn. London: E. Arnold.
13. Alexanderson, J. (1973) The influence of the properties of cement and aggregate on the consistence of concrete, in *Fresh Concrete: Important Properties and their Measurement, Proceedings of a RILEM Symposium 22–24 March 1973*, **2**, 3.2-1–3.2-16. Leeds, The University.
14. Teychenné, D.C., Franklin, R.F. and Erntroy, H.C. (1988) *Design of Normal Concrete Mixes*, revised edn. Building Research Establishment.
15. Tattersall, G.H. (1976) *The Workability of Concrete*, Viewpoint Publication, Cement & Concrete Association.
16. Wierig, H.-J. (1984) *Verfahren zur Prüfung der Konsistenz von Frischmörtel und Frischbeton*. Schriftenreihe des Bundesverbandes der Deutschen Transportbetonindustrie.

2 Standard tests for workability

In his review of workability tests, already mentioned, Wierig indicates which of them have been adopted as standard in 24 different countries, by means of a table, a slightly compressed version of which is given here as Table 2.1. Although the information is not completely up to date (e.g. the flow test has been withdrawn from the ASTM specifications in the USA), Wierig's table shows very clearly the emphasis laid by the various authorities. There is almost complete unanimity in selection of the slump test and there is strong support for a flow test (usually a flow table), the Vebe consistometer, and a compaction test, so these will be discussed in detail.

Most national standards specify several different tests and the reason for this, of course, is that no one of the tests is capable of dealing with the whole range of workabilities that is of interest in practice. Sometimes the range for which a particular test is considered to be suitable is stated; for example, in BS 1881 the recommendations shown in Table 2.2 are given. Developments in concrete technology may result in the addition of further tests at intervals; for example, for many decades BS 1881 contained three tests, slump, compacting factor and Vebe, but since none of these is capable of dealing with very high workability, the advent of flowing concretes associated with the use of superplasticizers resulted in the addition of the flow-table test taken from the German DIN specification.

The inclusion of a test in a standard does not necessarily mean that it is actually used in practice. In the UK, the Vebe consistometer is rarely (if ever) used outside a laboratory and it is probably true to say that most practising engineers have never even heard of it. By far the most commonly used test, in specifications and in practice, is the slump test; less frequently, the compacting factor is used for lower-workability concretes of the pavement quality type.

Table 2.1 Consistency test procedures standardized in various countries (shortened version of table given by Wierig[16])

	Country	Slump	Flow			Remoulding		Compaction		Penetration
			DIN	ASTM	Other	Vebe	Other	DIN	BS	
	ISO	x				x		x		
	RILEM	x				x				
A	Austria	x	x			x		x		
AUS	Australia	x							x	
B	Belgium	x	x		x	x				
CH	Switzerland	x	x			x		x		
CS	Czechoslovakia	x				x	x	x		
D	Germany (FRG)	x	x		x			x		x
DDR	Germany (GDR)	x	x			x	x	x		
E	Spain	x		x						
F	France	x		x						x
GB	Great Britain	x	x			x			x	
H	Hungary	x	x			x	x		x	
I	Italy	x		x						
IL	Israel	x		x						
J	Japan	x								
N	Norway	x				x	x			
NL	Netherlands	x	x					x		
PL	Poland	x				x				
P	Portugal	x		x		x				
S	Sweden	x			x	x	x			
SF	Finland	x	x		x	x	x			
SU	Soviet Union	x				x	x			
USA	USA	x		x						x
YU	Yugoslavia	x				x		x		
ZA	South Africa	x				x			x	

Table 2.2 Recommended ranges for test methods
as in BS 1881

Workability	Method
Very low	Vebe time
Low	Vebe time, compacting factor
Medium	Compacting factor, slump
High	Compacting factor, slump, flow
Very high	Flow

2.1 SAMPLING

Of course, it is a waste of time to carry out a control test on any material unless the sample is representative of the bulk, so if a standard were to fail to deal with this matter it would be deficient in an important respect.

Part 101 of BS 1881 lays down a method of sampling fresh concrete on site, and the later parts, dealing with the workability tests, require that the specified method be used. A standard scoop of approximately 5 kg capacity is specified and the number of scoopfuls needed for each of the tests is given (six for compacting factor, four for each of the others), each one to be taken from a different part of the batch. When sampling from a batch mixer or a ready-mixed concrete truck, the very first and very last parts of the discharge are to be disregarded, and, for the truck, the division into parts may be on the basis of a given number of discharging revolutions of the rotating drum. Sampling from a stream of concrete is to be done by passing the scoop through the whole width of the stream in a single operation.

Part 102 of the Standard permits a modification of sampling procedure for the measurement of slump of concrete delivered in a truck, in that the sample may be taken from the initial discharge after allowing a discharge of about $0.3 \, m^3$, but when this is done the sample should be split in half and a test carried out on each half. Because the sample cannot be regarded as fully representative, the permitted limits on the result are increased beyond those applying for those from samples obtained as in Part 101.

Once the sample has been obtained, the specification for each of the four BS workability tests requires that it be further homogenized by thorough mixing by hand on a tray and turning over three times.

The requirements of the ASTM Standard C 172-82 are similar to, though somewhat less detailed than, those of BS 1881. Sampling from either stationary or truck mixers is to be 'at two or more regularly spaced intervals during discharge of the middle portion of the batch' and a receptacle (not specified) is to be passed through the entire discharge stream. Individual samples are to be combined.

The justification for these sampling procedures will be discussed later in relation to mixing processes. There is no doubt that, in practice, it is common for inadequate attention to be paid to them. The *ASTM Manual of Aggregate and Concrete Testing* specifically warns against the development of unsatisfactory procedures when in Section 15 it refers to the standard C 172 and goes on to say: 'The practice of sampling concrete from the discharge stream of a mixer or truck, or from a pile, by means of a scoop or shovel, and then filling a test container (e.g. a slump cone) with several such samples without the required remixing should not be permitted'.

2.2 THE SLUMP TEST

As remarked earlier, the slump test is by far the best known and most commonly used test and is based on original proposals by Abrams. The apparatus as specified in BS 1881 Part 102 is shown in Figure 2.1. It consists of a hollow frustrum of a cone made of sheet metal and open at both ends (height 300 mm, bottom diameter 200 mm, top diameter 100 mm). It is held down firmly on a non-absorbent baseplate by the operator standing on the footpieces provided, and is filled from the top in a standard manner, i.e. in three layers with a standard method of rodding. After filling, the cone is lifted vertically ('slowly and carefully in 5 s to 10 s') and the concrete is allowed to slump. The amount of slump is measured and is recorded to the nearest 5 mm.

During the test the baseplate is supposed to be horizontal and free from shock, and it is suggested that this can be achieved by carefully levelling it on a bed of sand. In practice, on site, it is usually simply placed on the nearest patch of what seems to be reasonably level ground.

There are various types of slump as illustrated in Figure 2.2: the true slump, the collapse, and the shear slump, when one half of the cone shears off along an inclined plane. The Standard says that the test is valid only if it yields a true slump where the concrete remains substantially intact and symmetrical, and that if the specimen shears or collapses another sample must be taken and the test repeated. The report is required to state what form of slump was obtained and, if it was a true slump, the value of the slump as measured to the highest point of the slumped specimen. In practice it is quite common for what would otherwise be a true slump, as defined in the Standard, to be asymmetrical by leaning slightly to one side so that the top surface

Figure 2.1 The slump test: measuring the slump. (*Photo courtesey of the British Cement Association*).

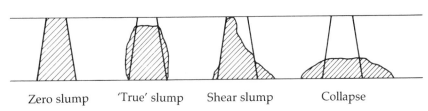

Zero slump   'True' slump   Shear slump   Collapse

Figure 2.2 Types of slump.

is no longer horizontal; in such case, measurement to the highest point would normally be reported.

There may be slight differences in apparatus or procedure in various countries and, for example, the cone for the slump test as specified in ASTM specification C 143-78 has a height of 12 in, a bottom diameter of 8 in and a top diameter of 4 in. These dimensions are the same as those that applied to the British cone before the UK adoption of SI units. There is no reference in C 143 to a required symmetry of the slumped cone but, as in the British test, shear slump results are to be discarded. The *ASTM Manual*, quoted above, refers in Section 18 to the slump test and adds the warning that rotation of the metal cone as it is withdrawn is not allowed.

The slump test is sensitive to variations in operator technique, intended or otherwise, and is also subject to other criticisms summarized, for example, by Glanville, Collins and Matthews[1] 40 years ago. They found that, for one particular mix, the addition of a small amount of water would result in collapse or shear slumps, while only a little less water would give a much more uniform slump of less than 25 mm. On the other hand, similar slump results could be obtained from mixes of different observed workabilities. The test is, of course, quite incapable of differentiating between concretes of different levels of very low workability, which all give zero slump, or of different levels of very high workability, which all give collapse slump.

2.3 THE COMPACTING-FACTOR TEST

The compacting-factor test was devised by Glanville, Collins and Matthews[1] because they recognised the importance of achieving full compaction in concrete, and therefore the importance of being able to measure the ability of the material to be compacted. They argued that the work in placing concrete is composed of that lost in shock, and the useful work, which is expended in overcoming the internal friction of the concrete itself and in overcoming the friction against the mould and the reinforcement. Of these, it is only the loss against internal friction that is characteristic of the concrete alone and it is this that they used as the basis for a definition of workability, and that they set out to measure.

They found that it was impractical to measure the work required to produce a given degree of compaction, so they finally developed the compacting-factor apparatus in which the inverse quantity, the degree of compaction produced by a given amount of work, is measured instead. The standard quantity of work is provided simply by allowing the concrete to fall under gravity through a standard distance.

The apparatus, shown in Figure 2.3, consists simply of two conical hoppers and a cylindrical mould mounted vertically one above the other, the capacity of the top hopper being greater than that of the lower, which in turn is greater than that of the cylinder. The internal surfaces are smooth, to minimize surface friction.

In performing a test, the top hopper is filled with concrete, using a scoop, and then the door at the bottom of the hopper is opened to allow the concrete to fall into the lower hopper which it fills to overflowing. The door of the lower hopper is then opened and the concrete falls into the cylinder and fills that to overflowing. After the excess has been cut off with two steel floats, the mass of the partially compacted

Figure 2.3 The compacting-factor apparatus where concrete is dropped from the first into the second hopper and then into the cylinder. (*Photo courtesey of the British Cement Association*).

concrete is determined by weighing to the nearest 10 g, and is divided by the mass of concrete required to fill the cylinder at full compaction to give a result known as the **compacting factor** which is, of course, a density ratio. The mass for full compaction is obtained by filling the cylinder with concrete on which a sufficient amount of work has been done either manually by rodding, or by the use of a vibrating hammer or vibrating table. The compacting factor is to be reported to two decimal places.

The test suffers from the disadvantage that a cohesive concrete tends to stick in the hopper and must be encouraged to fall by pushing a rod through it. This is particularly so for air-entrained concretes.

Cusens[2] showed that the amount of energy imparted to the concrete in the compacting-factor test is much less than that used in compacting a concrete by vibration. His results are shown in Figure 2.4, from which it can be seen that, for example, a mix whose compacting factor (or density ratio) was 0.70, achieved a density ratio of 0.88 when it was vibrated under the very mild conditions of 1.5 g at 100 Hz. Cusens concluded, therefore, that the test seems unsatisfactory for dry mixes, and Hughes[3] believes that, when vibration is used, the Vebe consistometer is preferable.

Although it is not strictly in accord with the requirements of the Standard, the mass of fully compacted concrete may be found by

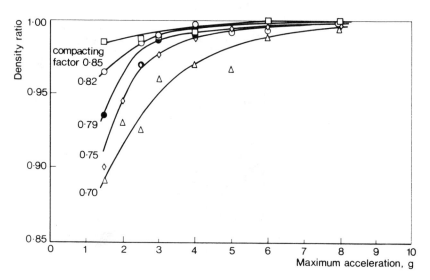

Figure 2.4 Relationship between density ratio and maximum acceleration determined from tests on 102 mm cubes. All specimens vibrated for one minute at 100 Hz. (*Cusens*)

compacting the partially compacted material using an internal vibrator, or poker, and adding further concrete. Anyone who has used this method must have noticed that a good estimate of the compacting factor may be made by noting the drop in level of the partially compacted concrete as it is vibrated and further compacted. A method of assessing compactability volumetrically is included in the German DIN standard, and one is also the subject of a British Draft for Development (or DD), so it might eventually be incorporated in BS 1881.

The German method, described in DIN 1048, consists of placing the concrete loosely in a metal box that is open at the top and has dimensions of height 400 mm, base 200 × 200 mm, and measuring the drop in height of the concrete as it is compacted. A compaction index is defined as the ratio of the initial height (400 mm) to the final height, and it should be noted that this is a quantity that is inversely related to compacting factor.

2.4 THE VEBE CONSISTOMETER

The Vebe consistometer is the third of the British Standard tests. It is a development of the remoulding test of Powers[4], and was originally proposed by Bahrner[5] to provide a test meant to be of particular use for concretes intended for placing by vibration. The apparatus is shown in Figure 2.5. A cylindrical container (diameter 240 mm, height 200 mm) is mounted on a vibrating table (frequency 50 Hz, amplitude without concrete 0.35 mm) and a separate slump cone is provided. To carry out a test, a slump cone of concrete is formed inside the cylinder by filling the metal slump cone in a standard manner and then removing it. A transparent circular plate, of standard weight, whose diameter (230 mm) is a little less than the internal diameter of the cylinder, and which is mounted horizontally on the lower end of a vertical rod that slides in a guide, is placed on top of the concrete and the vibrator is switched on.

The effect of the vibration is to cause the concrete to subside and remould to the shape of the cylinder, a process which is observed through the transparent disc. The end point of the test is defined as the moment when the transparent disc is completely coated on its underside with cement grout, and the time required for this to happen is to be reported as the Vebe time in seconds, quoted to the nearest second.

If the slump occurring before the vibrator is switched on is a 'true' slump it is to be measured, but if it is a 'shear' slump or if the concrete subsides to touch the side of the cylindrical container, measurement of the Vebe time is to proceed straight away.

Figure 2.5 The Vebe apparatus. (*Photo courtesey of the British Cement Association*).

In practice, the vibrating table of the Vebe consistometer is activated by the rotating eccentric type of vibrator, so the waveform is much more complicated than a pure sine wave, and consequently the definition of the vibration parameters, frequency and amplitude, is not as simple as is implied in the Standard. However, the main criticism specific to this test is that the start is ill-defined because the vibration takes time to build up, and the end-point is ill-defined because the rate of wetting of the disc with grout decreases with time and may even reach zero before the whole area is covered. It is in principle bad practice to use an end-point that is approached at a decreasing rate, or even asymptotically.

Some investigators[6,7] have suggested the use of settlement/time recorders, but Hughes and Bahramian[8] found that the resulting curve did not facilitate more accurate assessment of the Vebe time. They did however suggest that the area under the curve could be used to give an indication of the cohesiveness of the concrete.

Bahrner introduced a correction factor to the Vebe time, t, and expressed his result in Vebe degrees given by $(V'/V_0)t$ where V_0 and V' are the volumes of concrete before and after vibration. Hughes and Bahramian have pointed out that since a concrete that loses its air voids in the making of the slump cone will have no correction applied to the Vebe time, whereas a less workable concrete that does not lose air until vibration is applied will have its Vebe time reduced, the effect of the correction factor is the opposite to what might be expected. They concluded that the use of Vebe degrees instead of Vebe time is an unnecessary complication.

2.5 THE FLOW-TABLE TEST

The advent of the use of superplasticizers to produce concretes of very high workability led to problems of assessment because it was obvious that none of the existing three British Standard tests could be used. The slump test could not cope because all these concretes gave a 'collapse' slump, and since the other two tests were developed for concretes of lower workability than those for which the slump test was thought to be appropriate, they could not cope either. The solution adopted was to include in the 1983 edition of BS 1881 a test taken from the German specification DIN 1048. This is the flow-table test.

The flow table, shown in Figure 2.6, usually consists of a wooden board 700 mm × 700 mm which is connected by hinges to a baseboard of similar dimensions. Thus when the baseboard rests on the floor the upper board can be raised through an angle to the horizontal which may be fixed at any chosen value by the provision of a suitable stop. In fact the stop is so arranged that the vertical movement of the free edge is limited to 40 ± 1 mm. The upper board, which has a total mass of 16 ± 1 kg, is covered on its upper surface by a metal plate of thickness not less than 1.5 mm, whose centre is marked by a cross, the arms of which are parallel to the sides of the plate and extend the full width and length. A concentric circle, 200 mm in diameter, is also marked. The front edge of the upper board is provided with a handle for lifting and the front of the lower board is extended not less than 120 mm beyond the front of the upper board, to provide a toe board.

In use, the flow table should be placed on a firm and horizontal level surface so that it does not rock, and the metal surface should be

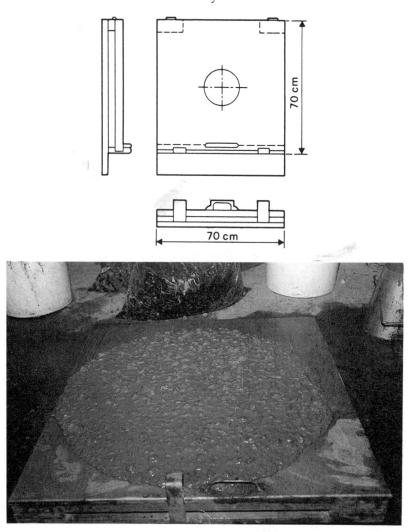

Figure 2.6 The flow table consists of a wooden board, faced with sheet metal, connected to a baseboard. (a) Diagram of the apparatus. (b) Flow table test. (*Photo courtesey of Tilcon Ltd*).

wetted. A truncated cone of the concrete to be tested is formed in the centre of the metal surface with the aid of a mould, similar to the normal slump cone mould in shape and in being provided with foot-pieces, but of height 200 mm and having upper and lower diameters of 130 mm and 200 mm. The mould is filled in two equal layers while the tester stands on the footpieces, and each layer is tamped ten times with a wooden tamper of base area 40 mm square. When the cone is full, the excess concrete is struck off with a steel straight edge and the plate is cleaned. Half a minute after striking off, the metal cone is removed by lifting slowly and vertically over a period of three to six seconds. After this, the upper board is lifted to the stop, without striking it heavily, released, and allowed to fall. This is done fifteen times, and each cycle is to take between three and five seconds.

The bumping action causes the concrete to spread, and after 15 bumps the two diameters parallel to the sides of the table are measured. The 'spread' is taken as the arithmetic mean of the two diameters, expressed to the nearest 5 mm.

This test was severely criticized by Dimond and Bloomer[9] some six years before it was incorporated in the British Standard, and they recommended that its use '. . . should cease before a bank of useless and misleading information has been built up.' Their reasons were as follows:

(a) It is likely to be operator-sensitive mainly because of uncertainties about the force with which the limiting stop is struck when the table is raised.

(b) Simple calculation shows that, after the concrete has spread as a disc to a diameter of 51 cm, which is recommended as the minimum final spread for a concrete to be classified as a flowing concrete[10], its thickness is only about the same as the dimensions of a normal 20 mm aggregate. Thus there is no hope that the test can measure bulk properties.

(c) Laboratory tests on a range of high-workability mixes showed a highly significant correlation ($r = 0.954$) between the final spread S_f, measured as in the Standard, and the initial spread S_0, measured immediately after the slump cone had been removed. This is shown in Figure 2.7 and suggests that the initial spread is at least as good (or as bad) a measure of workability as final spread. Of course, measurement of initial spread is essentially only an alternative way of carrying out a slump test and is therefore subject to all the criticisms that apply to that.

(d) For four different high-workability mixes the spread was measured after every bump and the results were as shown in Figure 2.8. It can be seen that the lines approach each other as the number of bumps increases so that the difference in spreads is reduced by bumping.

(e) Although it has been argued correctly that the test will indicate whether the concrete is likely to segregate, additional subjective visual judgment is required. That judgment can be made equally well on a concrete that has simply been tipped on the floor.

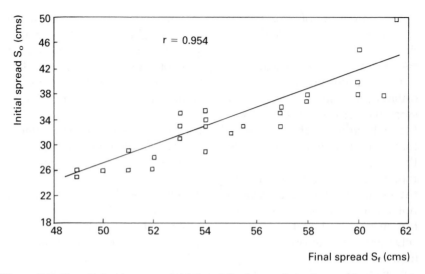

Figure 2.7 Correlation between initial and final spreads in flow-table tests. (*Dimond and Bloomer*)

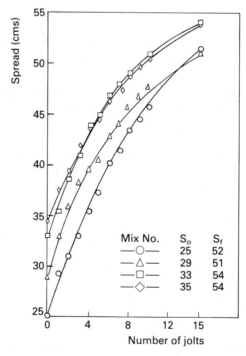

Figure 2.8 Relationship between spread and number of jolts in flow-table tests for four high-workability concretes. (*Dimond and Bloomer*)

The criticisms of Dimond and Bloomer have in their turn been criticized by other workers. Mor and Ravina[11] consider that because the operation (of bumping) takes place 15 times there is a reduction in the probability of random operator error. Even if this is so, it does not reduce the probability of differences between operators caused by differences in interpretation of the requirement to lift the board to the stop 'without striking it heavily', and that is the point Dimond and Bloomer were emphasizing. Mor and Ravina also express the opinion that although the thickness of the concrete will be of the same order as the maximum aggregate size, 'the concrete's behaviour will represent the combination of its components', but they do not state clearly that the behaviour they refer to here is behaviour **in the test**, which may not relate in any meaningful way to behaviour under circumstances (i.e. in practice on site) where the bulk properties are important.

Dimond and Bloomer's finding of a correlation between initial and final spreads has been questioned by Dhir and Yap[12] who worked on a wider range of concrete workabilities and produced the results shown in Figure 2.9. Here there is a distinct turnover in the curve at lower workabilities. However, an initial spread of 200 mm means that the concrete has not moved at all before bumping, because the mould

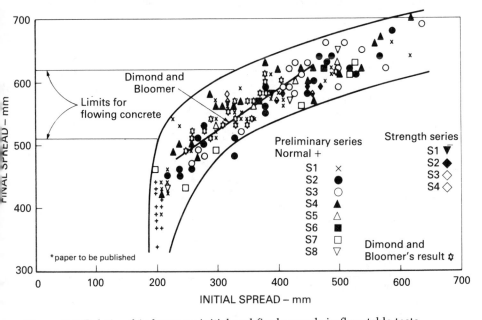

Figure 2.9 Relationship between initial and final spreads in flow-table tests. (*Dhir and Yap*)

diameter is 200 mm. If the points for concretes with this initial spread of 200 mm are removed from Figure 2.9, the remaining points show a simple linear relationship but with a wide spread about the best line. This suggests the making of a fairly arbitrary choice between the two possible measures, initial spread or final spread. The results of Dhir and Yap do show that bumping can provide some sort of differentiation between concretes that all give the same initial spread of 200 mm (i.e. that have not moved at all in the absence of bumping), and it might therefore be claimed that final spread is the more sensitive indicator. However, this does not apply at the other end of the workability range as shown by the results of Dimond and Bloomer already discussed (Figure 2.8).

The finding that there is a correlation between initial and final spreads for high-workability concretes, together with the statement that measurement of initial spread is only another way of measuring slump, receives support from results of workers who have investigated the relationship between final spread and slump measured in the normal way. Mor and Ravina quote results to this effect from the work of Hewlett[13] and Meyer[14] and they themselves carried out experiments on 42 gap-graded concretes, many incorporating superplasticizers and with slumps from 110 to 250 mm. Their results are shown in Figure 2.10. They quote a correlation coefficient of 0.94 and give equations

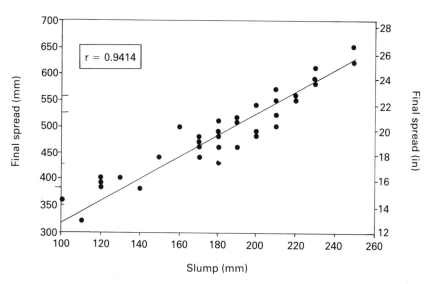

Figure 2.10 Relationship between final spread and slump. (*Mor and Ravina*)

which they say can be used to translate results from one system to the other. If S and F are the slump and flow values in inches they say,

$$S = 0.5 F - 2.17$$
$$F = 2 S + 4.3 \tag{2.1}$$

Notice that the second of these equations is only an alternative way of stating the first: the two equations are one and the same. It is not stated how the equation was obtained, but the best equation for predicting S from F is not the same as the best equation for predicting F from S; two different equations should have been obtained from the results.

Finally, Moh and Ravina believe that Dimond and Bloomer's finding that final spread was less sensitive than initial spread for differentiating between concretes is not important because: 'It can be deduced that concrete mixes with different static behaviour showed their real dynamic behaviour only after being bumped 15 times. They then exhibited similar dynamic properties.' Far from being a justification for the flow-table test, this is in fact a criticism of it, as will be shown in later discussion.

It may be concluded that the contentions of Dimond and Bloomer have, in spite of later criticism, been fairly well validated.

2.6 BALL-PENETRATION TEST

As the name implies, the general idea behind a penetration test is to measure the penetration into the fresh concrete of an element of a specified shape under a specified load, and this latter is often the self-weight of the element. Many different proposals have been made involving the use of elements of a variety of shapes, such as rod, sphere, or cone, but they have suffered a marked lack of recognition. One exception is the ball-penetration test that is the subject of ASTM Standard C 360-82, and the apparatus for which is shown in Figure 2.11. The penetrating element is in the form of a cylinder of total height 4 in and diameter 6 in, with its bottom machined to a hemisphere of 6 in diameter. The total mass is 30 ± 0.1 lb (13.6 kg) and a guiding frame is provided as shown. In use, the apparatus is placed on the surface of the concrete, which must be smooth and level, and the depth of penetration after release is measured to the nearest $\frac{1}{4}$ in (6.4 mm). At least three readings at positions not less than 6 in apart are to be taken, then if the difference between the maximum and minimum readings exceeds 1 in, more readings are to be obtained until three successive ones agree to within 1 in. The mean of three is to be reported.

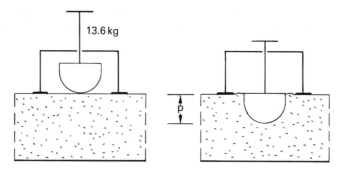

Figure 2.11 Ball-penetration test (radius of hemisphere 3 in [76 mm]).

A practical advantage of this test is that it can be carried out on concrete in any suitable container or even in formwork and it seems to be intended as a substitute for, or a supplement to, the slump test, because Clause 4.1 says: 'After sufficient correlation data with results from the standard slump test is (sic) obtained the results of the penetration reading may be used to determine compliance with slump requirements.'

2.7 PRECISION OF THE STANDARD TESTS

A measurement of any physical quantity is of very limited value unless an estimated experimental error can be assigned to it, that is, unless it is known how accurate and reliable is the figure quoted. It is highly desirable to be able to calculate an error for every separate experimental result, as is normal practice in many areas of measurement, but that cannot be done for any of the standard workability tests, so the somewhat less satisfactory alternative of quoting an overall figure for the particular test method must be considered.

In this connection, BS 1881 cross-refers to another British Standard, BS 5497 Part 1: 1987, entitled *Precision of test methods. Part 1. Guide for the determination of repeatability and reproducibility for a standard test method*. This latter Standard defines two quantities, r and R, which are respectively the repeatability and the reproducibility. The repeatability, r, is defined as: 'The value below which the absolute difference between two single test results obtained with the same method on identical test material under the same conditions (same operator, same apparatus, same laboratory, and a short interval of time) may be expected to lie with a specified probability; in the absence of other indications the probability is 95%'. The reproducibility, R, is defined in a similar way except that it refers to different conditions, that is, to results obtained by different operators with different pieces of apparatus, in different laboratories or at different times. The difficulty of sending 'identical test material' to different laboratories, when that material is fresh concrete, is obvious, so if R is to be investigated it is necessary to bring all the participating 'laboratories' to the same site at one time.

The ASTM Standard C 670-87 *Preparing precision and bias statements for test methods for construction materials* refers to a quantity named as 'the difference two-sigma limit (D2S)' which is to be obtained for single-operator and for multi-laboratory results. The two figures so obtained are exactly the same as the r and R of the British Standard but the ASTM view is '. . . because of considerable confusion concerning the exact meaning of those terms (i.e. repeatability and reproducibility) their use is not recommended'. Attention is drawn to the difference between laboratory and field conditions, a point that does not seem to be made specifically in the British Standard, and it is recommended that appropriate studies be made 'for tests under field conditions'. No statement about precision of the slump test is made by ASTM and for the ball-penetration test it is stated that precision has not yet been determined but is being investigated.

In the parts of the British Standard, BS 1881, dealing with the compacting factor, the Vebe consistometer, and the flow table, it is specifically stated 'No estimate of repeatability or reproducibility is given in this Part of this British Standard'. In Part 102, which deals with the slump test, no estimate of reproducibility is given, but it is stated that 'For slump measurements made on concrete taken from the same sample, the repeatability is 15 mm at the 95% probability level, for normal concrete having a measured slump within the range of 50 mm to 75 mm'. Evidence to justify the statement is not given.

Estimates of repeatability may be made from the figures given in Table 2.3 which shows results quoted by Banfill[15] from factorial experiments by Orr[16] for compacting factor, and by himself for slump

Table 2.3 Variability of BS 1881 tests

Test	Range	Standard deviation	Degrees of freedom	Dependence on response
Slump	0–100 mm	11 mm	21	Independent
CF	0.75–1.00	0.024	60	Independent
Vebe	1–24 s	$0.25 \times V$	23	Coefficient of variation constant

and Vebe time. The repeatability, r, at the 95% probability level, for single tests is obtained by multiplying the standard deviation by 2.8, so for the slump test it is approximately 30 mm. This is double the figure given in BS 1881.

Work has recently been carried out specifically with the objective of providing figures for the Standard, and the results have been reported by Sym[17]. Sixteen operators were involved in the experiment, eight of them in each of two sessions. In each session a 0.2 m³ batch of concrete was made and half of it discharged into a 0.1 m³ pan from which the operators sampled to carry out the first set of tests. About 20 minutes later they were presented with the second half of the batch and carried out the repeat tests. In the cases of compacting factor and slump the whole experiment was repeated with a second batch (although the mixes were not the same for the two tests) but for the flow table only one batch was used. Sym gives values of r and R for both single tests and for the mean of the results of two tests, but discussion here will be restricted to the single tests as being of most interest for practical site conditions. The figures given, which are shown in Table 2.4, were calculated by means of the split-level method described in BS 5497. These figures have been used to prepare amendments to BS 1881 which will be issued in the near future[18].

Several comments may be made about these results. In the case of slump, the original figures on which the results are based do not seem to provide any justification for quoting different figures for two slump mean values of 80 mm and 50 mm, but, more importantly, a whole block of figures (one quarter of all results), relating to a particular mix in one particular session, was rejected and omitted from further consideration because the concrete 'was excessively workable and segregated to some extent, so that large differences were observed between the first and second tests. The repeatability was noticeably worse with this batch than for the others'. No other reasons are given and the propriety of neglecting results on these grounds, after the material has been accepted and the tests carried out, may be questioned, particularly since the object of the work was to investigate

Table 2.4 Precision results reported by Sym

Test	r	R
Compacting factor	0.031	0.040
Slump (80 mm)	16 mm	30 mm
(50 mm)	17 mm	23 mm
Flow	69 mm	91 mm

repeatability and reproducibility. Examination of the rejected results suggests that it might be justified to quote separately higher values of r and R for higher slumps (of the order of 100 mm), but if this is not done and the results are included with the others the effect is to cause a substantial increase in the estimated value of r (to about 21 mm) and a lesser increase in R.

In practice, the value of R is likely to be of more interest than the value of r, because the former is concerned with the agreement between results obtained by different operators and the latter with that between results obtained by the same operator. The usual practical problem concerns comparing a slump value found on site with the value that the supplier says it is; that is, comparing values from two **different** sources. Overall, considering the results of both Sym and Banfill, it seems that a reasonable figure to take for R is 30 mm, with a corresponding standard deviation of 11 mm.

For compacting factor, there is a substantial difference between the value of R of 0.040 given by Sym and the value of 0.067 that may be calculated from Orr's results; perhaps all that can be suggested at present is to take a value somewhere in between, say 0.05 or 0.055, with a corresponding standard deviation of about 0.02. Further work needs to be done on this subject.

In any case, in interpreting these results it should be remembered that they were obtained from laboratory tests, and not under site conditions. Sym states: 'The operators were in most cases, experienced with concrete testing, and were also well aware that they were taking part in a special event, so one would expect the precision results achieved in this experiment to represent a better-than-average standard of testing.'

Another important point is that in the analysis carried out by Sym the effect of time lapse between testing of duplicate samples was eliminated by assuming that, for a given batch of concrete, the effect was the same for all operators. This means that the results should properly be applied only to two tests that are carried out simultaneously or within a short interval of time. In practice, this may not be possible and, if it is not, some allowance for change in workability with time must be made, although that is not easy to do. The effects of time lapse on workability are discussed in Chapter 8.

2.8 GENERAL CRITICISM OF STANDARD TESTS

Criticisms have already been given of each of the standard tests individually and further problems will appear when their use in connection with specification of concrete is considered. There are other, and more serious, criticisms that may be applied to all of them as a class.

Each of the tests can cope with only a limited range of workabilities and, for example, BS 1881 lays down the permitted limits for use as in Table 2.5. These limitations would not necessarily be important if all the tests measured the same property; nobody finds it odd that many different forms of instrument are used in the measurement of length, and that, where a micrometer is appropriate, a builder's tape would be useless. However, whereas the difference between a micrometer and a tape lies only in their accuracy, so that they both measure the same property in the same scale of units, the various workability tests all measure different things on scales that cannot be related to each other satisfactorily.

For example, Dewar[19] showed that the relationship between compacting factor and either slump or Vebe time was significantly affected by the type of fine aggregate and the richness of the mix, and Hughes and Bahramian[8] provide similar evidence. In each of the relevant parts of BS 1881 it is specifically stated that: 'There are no unique relationships between the values yielded by the four tests. Relationships depend upon such factors as the shape of the aggregate, the sand fraction and the presence of entrained air'.

The fact that the range of each test is limited, coupled with the absence of relationships between the results they give, means of course that there is no hope of using them to establish a workability scale into which they would all fit and which would cover the whole range of practical workabilities.

Further, it is very important to realize that the recognition, clearly stated in the British Standard, that there are no unique relationships, is a tacit admission that the tests are not all measuring the same property and, in fact, that none of them is measuring workability. This tacit admission is somewhat more explicitly stated in Part 102 of BS 1881 where a footnote advises: 'Some indication of the cohesiveness and workability of the mix can be obtained if, after the slump measurement has been completed, the side of the concrete is tapped gently with the

Table 2.5 Ranges of suitability of BS 1881 tests

Test	Permitted limits
Slump	5–175 mm
CF	0.70–0.98
Vebe	3–30 s
Flow	None stated

tamping rod. A well-proportioned concrete which has an appreciable slump will gradually slump further, but a badly proportioned mix is likely to fall apart'. Similarly, a footnote in Part 105, referring to the flow-table test, says: 'The concrete spread may also be checked for segregation. The cement paste may segregate from the coarse aggregate to give a ring of paste extending several millimetres beyond the coarse aggregate.' When Glanville, Collins and Matthews[1] first described the compacting factor test, they recognized that it was not a sufficient measurement for assessing workability and they supplemented it with a slab test and a heap test. The former consisted of filling a slab mould in a standard manner and examining the surface of the resulting slab for honey-combing, while the latter consisted simply of photographing a heap of concrete that had been dropped in a standard way.

Associated with the fact that they do not measure workability is the most damning practical defect of all, and that is that any one of the standard tests is capable of classifying as identical in properties two concretes that are subsequently found to behave very differently in practice. For example, it is well known that two concretes of the same slump may be found to have quite different workabilities on the job. This most serious fault, which means that these tests are not even suitable as simple pass/fail tests, can have awkward and very expensive consequences in practice, because major errors in batching can remain undetected until results of cube tests become available. Examples will be given later in the discussion on quality control.

This defect of the standard tests, that is, their inability to distinguish between concretes of different observed workabilities, is not considered explicitly in any of the standards. BS 5497, which deals with precision of measurements, is concerned only with the important question of ensuring that a test gives identical results on identical samples, within acceptable experimental errors, and does not deal at all with the equally important question of ensuring that a test gives different results on different samples. It may be remarked here that it is possible to obtain any desired measure of agreement between results of a given test simply by making the test sufficiently insensitive.

On top of all this, there is another quite serious deficiency of the standard tests, and that is that they are incapable of giving any indication of the cause of any unwanted change in workability. If the slump value of a concrete is not within the specified range there is no way of telling from the results of the slump test why it is not. In practice, if a slump value is considerably higher than it should be, it is likely that it will be assumed that the reason is that the batch contains too much water. In one particular case, concrete with a slump higher than specified was delivered to the job and placed in a column. When the Resident Engineer discovered this later, he insisted that the column be demolished, although non-destructive testing of the hardened concrete indicated that the suspect material was as strong as the rest. His thinking was based on the quite erroneous assumption that the higher workability **must** have been due to a higher water content, and as a result he made a decision that was quite unreasonable.

These general criticisms of the standard tests apply with equal force to all the other empirical tests, and while it would be foolish to deny that such tests have assisted in matters like the development of methods of mix design, there are strong indications that their day is now over. They may have been appropriate in the early days of concrete technology but they are not so in the light of more advanced knowledge and when there is increasing emphasis on quality assurance. Something better is very much needed.

2.9 REFERENCES

1. Glanville, W.H., Collins A.R. and Matthews D.D. (1947) *The grading of aggregates and the workability of concrete*, 2nd Edn. Road Research Technical Paper No. 5. London, HMSO.
2. Cusens, A.R. (1956) The measurement of the workability of dry concrete mixes, *Magazine of Concrete Research*, **8**(22), 23–30.
3. Hughes, B.P. (1972) The economic utilization of concrete materials, *Proceedings of a Symposium on Advances in Concrete, Birmingham 1971*. London, Concrete Society. 1972, p. 18.
4. Powers, T.C. (1932) Studies of workability of concrete, *Journal of the American Concrete Institute, Proceedings*, **28**, Feb. 1932, pp. 419–48; June 1932, pp. 693–708.
5. Bahrner, V. (1940) The Vebe consistency testing apparatus, *Zement*, **29**(9), 102–6; *Betong*, **25**(1), 27–38.
6. Erntroy, H. (1956) Discussion of Reference 2, *Magazine of Concrete Research*, **8**(24), 184–5.
7. Meyer, U.T. (1962) Measurement of the workability of concrete, *Journal of the American Concrete Institute, Proceedings*, **59**(8), 1071–80.
8. Hughes, B.P. and Bahramian, B. (1967) Workability of concrete: a comparison of existing tests, *Journal of Materials*, **2**(3), 519–36.
9. Dimond, C.R. and Bloomer, S.J. (1977) A consideration of the DIN flow table, *Concrete*, **11**(12), 29–30 (discussion *ibid*. (1978) **12**(2), 18.)
10. Hewlett, P.C. (Editor). (1976) *Superplasticizing admixtures in concrete*. Report of a joint working party of the Cement Admixtures Association and the Cement and Concrete Association. Slough, Cement and Concrete Association, Publication 45.030.
11. Mor, A. and Ravina, D. (1986) The DIN flow table, *Concrete International*, Dec. 1986, 53–6.
12. Dhir, R.K. and Yap, A.W.F. (1983) Superplasticized high workability concrete: some properties in the fresh and hardened states, *Magazine of Concrete Research*, **35**(125), Dec. 1983, 214–28.

13. Hewlett, P.C. (1979) The concept of superplasticized concrete, in *Superplasticizers in concrete*, SP-62, American Concrete Institute, Detroit, 1–20.
14. Meyer, A. (1979) Experiences in the use of superplasticizers in Germany, in *Superplasticizers in concrete*, SP-62, American Concrete Institute, Detroit, 21–36.
15. Banfill, P.F.G. (1977) Discussion of Reference (7) Chapter 6. *Magazine of Concrete Research*, **29**(100), 156–7.
16. Orr, D.M.F. (1972) Factorial experiments in concrete research, *Journal of the American Concrete Institute, Proceedings*, **69**(10), 619–24.
17. Sym, R. (1988) Precision of BS 1881 concrete tests. Part 1: Data and analysis. British Cement Association, April 1988, 40pp (BSI Document 88/11780, Committee Ref. CAB/4).
18. Mears, A.R. Private communication to G.H. Tattersall, 24 January 1990.
19. Dewar, J.D. (1964) Relations between various workability control tests for ready-mixed concrete. London, Cement & Concrete Association, 1964. 17 pp. Technical Report 42.375.

Standards referred to

British

BS 1881: 1983 *Testing Concrete*
　　　　Part 101 Method of sampling fresh concrete on site
　　　　Part 102 Method for determination of slump
　　　　Part 103 Method for determination of compacting factor
　　　　Part 104 Method for determination of Vebe time
　　　　Part 105 Method for determination of flow
BS 5497: 1979 *Precision of test methods*
　　　　Part 1 Guide for the determination of repeatability and reproducibility for a standard test method

American

ASTM C172-82 *Standard method of sampling fresh concrete*
ASTM C143-78 *Slump test*
ASTM C360-82 *Ball penetration in fresh Portland cement concrete*
ASTM C670-87 *Preparing precision and bias statements for test methods for construction naterials*
ASTM *Manual of Aggregate and Concrete Testing*

German

DIN 1048 Teil 1 (Jen. 1972) *Prüfverfahren für Beton – Frischbeton, Festbeton gesondert hergestellter Probekorper*

International

ISO 4109 *Fresh concrete – Determination of the consistency – Slump test*
ISO 4110 *Fresh concrete – Determination of the consistency – Vebe test*
ISO 4111 *Fresh concrete – Determination of consistency – Degree of Compact-
ability (Compaction index)*
ISO 9812 *Fresh Concrete – Determination of consistency – Flow table*
Note: The ISO slump and Vebe tests are the same as the BS 1881 tests.
The ISO compaction test is the same as the DIN 1048 test.

Note on European standardization

It seems likely that the ISO tests will be accepted for use in European
standards.

3 Flow properties of fresh concrete

The fact that any of the standard tests, or the other empirical tests, can classify as being identical in flow properties two concretes that are subsequently found to behave very differently on the job, is very important indeed, because it means that none of these tests is satisfactory even as a simple pass/fail test. If an attempt is made to use one of them in this role it is quite probable that a concrete that is unsatisfactory will pass or, conversely, one that is satisfactory will fail. Clearly then, these methods do not measure any property of the fresh concrete that can be simply related to its behaviour on the job, so if any attempt is to be made to set up an adequate quality control system something better is needed. It is necessary to ask what it is that should be measured and how it can be done.

To answer these questions means that it is necessary in turn to ask what it is that governs the way concrete flows. A good way to start is to consider first the way that very simple materials flow and then see whether concrete is any different and, if so, how.

If water, or a light oil, or any other simple liquid, is made to flow through an apparatus such as that shown in Figure 3.1 so that the rate of flow at a series of different constant pressures can be measured, it will be found that there is a simple linear relationship, of the type shown in the graph in Figure 3.2. The characteristics of this simple relationship are as follows.
(a) However low the pressure applied, some flow takes place.
(b) The rate of flow is simply proportional to the applied pressure, so that if the pressure is doubled, or halved, so is the flow rate.
Both these characteristics are covered by expressing the results in terms of the very simple equation

$$Q = K \cdot P \qquad (3.1)$$

Where Q is the flow rate expressed as a volume or mass per second and P is the applied pressure. The constant K will of course depend on

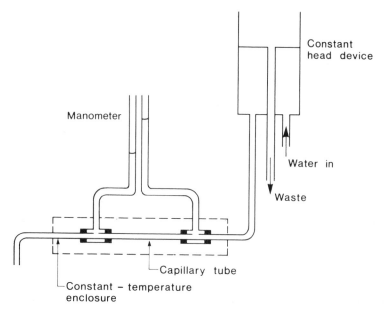

Figure 3.1 Principle of measurement of coefficient of viscosity of water by flow through a capillary tube.

the details of the apparatus, in particular on the length and internal diameter of the flow tube, but also, and very importantly it depends on the nature of the liquid under investigation. In fact, for a given experimental set-up, it depends only on the nature of that liquid and is a measure of its fluidity. If the experiment is repeated on a series of different liquids, it will be found that in each case a simple straight-line relationship is obtained, but the values of K will be different for the different liquids, which in turn means that a series of lines of different slopes will be obtained as shown in Figure 3.3. For each liquid, the corresponding value of K completely characterizes the behaviour of the particular material in this particular apparatus. In fact, it is an easy matter to eliminate the apparatus effect and to obtain from each value of K a value of the **fluidity** ϕ, or its reciprocal which is called the **viscosity** η, in fundamental units independent of the apparatus used to obtain the results. Equation 3.1 shows that since the liquid is completely characterized by the single constant K (or ϕ or η derived from it), only one measurement is needed to permit a value for K to be deduced; measurement of rate of flow at only one pressure is sufficient for substitution in the equation to allow K to be calculated. The

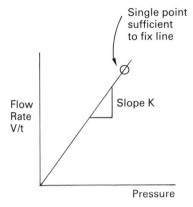

Figure 3.2 Relationship between flow rate and pressure for a simple liquid flowing through a tube.

sufficiency of a single experimental measurement can also be seen geometrically as in Figure 3.2; a single experimental point is enough to fix the position of the line, or flow curve, because there is only one straight line that will pass through that point and through the origin.

The experimental set-up described is essentially what is known as a **capillary-tube viscometer**. An alternative, and for several reasons a generally more useful form of apparatus, is the **coaxial-cylinders viscometer**, which is also known as a concentric-cylinders viscometer, a Couette viscometer, or sometimes simply as a rotation viscometer, although this last term includes other types as well. As the name implies, the essential part of this instrument consists of two cylinders which are mounted coaxially (Figure 3.4) in such a way that there is a space between them, the gap, which is filled with the material under

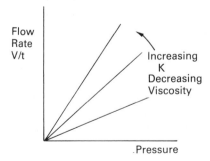

Figure 3.3 Relationship for liquids of different viscosities.

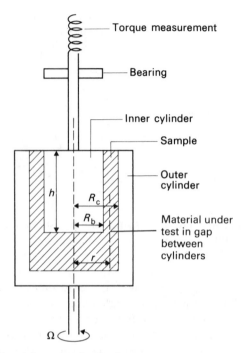

Figure 3.4 Principle of the coaxial-cylinders viscometer.

investigation. One of the cylinders, often the outer one, is rotated at a known speed with the result that the other cylinder also tends to turn because of the viscous drag acting on it. The torque required to prevent it from turning is measured.

If this apparatus is used to study a simple liquid of the type so far discussed it is found that there is a simple linear relationship between the torque on the stationary cylinder and the speed of the rotating cylinder as shown in Figure 3.5. Again, the slope of the line is a measure of the fluidity of the liquid, that is:

$$T = C \cdot N \tag{3.2}$$

Where T is the torque at a speed N rev/s and C, the proportionality constant, is the slope of the line. C is proportional to the fluidity ϕ and its reciprocal $1/C$ is proportional to the viscosity η. Again, it is a simple matter to eliminate the effects of apparatus dimensions and to obtain ϕ and η in absolute units.

These two linear results (equations 3.1 and 3.2) are obtained because the underlying relationship for these materials is itself a simple linear

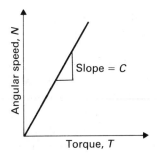

Figure 3.5 Relationship between torque and angular speed for a Newtonian liquid in a coaxial-cylinders viscometer.

one, that is, the shear stress is proportional to the shear rate, and the constant of proportionality is called the viscosity. This underlying law is called **Newton's Law of Viscous Flow**, and a substance that obeys it is referred to as a **Newtonian liquid**. It is a remarkably simple law. There is no a priori reason why it should be supposed that the relationship between shear stress and shear rate, or between pressure and flow rate in a tube, or between torque and angular speed in a coaxial-cylinders viscometer, should be so simple and in fact, for many materials of great practical importance it is not. The flow curve may not be linear, it may not start from the origin, and, worst of all, it may not be reproducible; in any of these cases a single constant will **not** be sufficient to characterize the material.

It is reasonable to suspect then that anyone who assumes, without checking, that fresh concrete behaves as in the simple linear case is pushing his luck, and need not really be surprised if proposals he makes based on that assumption turn out to be unsatisfactory. All the standard tests, and the other empirical tests, attempt to assess workability in terms of a single quantity, be it a slump value, a Vebe time, or any other, so they all involve the tacit assumption that concrete behaves in the simplest of all possible ways, like water, or a light oil.

The most casual observation of the behaviour of concrete shows that this assumption cannot possibly be true. First of all, it is obvious that the imposition of some minimum stress or force is necessary to get concrete to move at all, which means that it possesses what is known as a **yield value**, and consequently the flow curve cannot possibly pass through the origin. The mere fact that a slump test can be carried out at all, or that the material can stand in a pile and resist flow under the influence of its own self-weight, shows that this is so. The slump test does give a crude assessment of the yield value of a concrete but it gives no information at all about the extra forces that are needed as soon as the concrete is made to move at speeds that are important in practical jobs.

Thus, without doing any experiments at all, other than just looking at the material on site, it can be deduced that the flow curve does not pass through the origin and therefore cannot possibly be characterized by a single constant. To find out what the shape of the flow line is, it is necessary to carry out some experiments in which measurements are made at a series of different flow speeds. If concrete is examined in a suitably designed coaxial-cylinders viscometer it is found that the relationship between torque and speed is as shown in Figure 3.6: that is, it is a simple straight line but, unlike the Newtonian case, it does not pass through the origin; it has an intercept on the torque axis. The equation of this line may be written

$$T = C' + C'' \cdot N \tag{3.3}$$

Where T is the torque at angular speed N, as before. C'', the slope of the line, again is a measure of fluidity but in this case the term **mobility** is used; its reciprocal $1/C''$ is again a viscosity, which is called the **plastic viscosity**. The new feature is the constant C' which is equal to the intercept on the torque axis and is a measure of the yield value, that is, the torque at zero speed, or the torque that is the minimum to cause the concrete to move at all.

The fact that there are now two constants needed to describe the material and its flow curve means of course that it is necessary to obtain experimental results at not fewer than two different speeds so that two simultaneous equations of the form of equation 3.3 can be set up and solved for C' and C''. The same thing can be seen on a

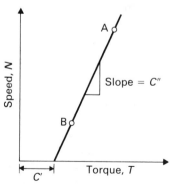

Figure 3.6 Relationship between torque and angular speed for Bingham material in a coaxial-cylinders viscometer.
A and B represent the two experimental points needed to fix the line.

geometrical basis in Figure 3.6; if only one experimental point is available, the flow curve might be any one of the infinite number of possible lines to pass through that point, and a second point is required to decide which of them is the correct one.

Equation 3.3 is a particular expression of an underlying fundamental relationship that is known as the **Bingham model** and says that the shear stress on a material is the sum of a yield value and another term that is proportional to the shear rate. The constant of proportionality in the second term is called the **plastic viscosity** and its reciprocal is called the **mobility**. A substance whose flow properties conform to this model is called a **Bingham material**. Concrete is therefore a Bingham material because as has been shown by a great deal of experimental evidence its flow properties conform quite closely to the Bingham model.

This is indeed a fortunate result from a practical point of view because it means that although concrete is not a simple Newtonian liquid and cannot be characterized by one constant, it is the next simplest case, a Bingham material, and can be described by only two constants. Thus, it seems that the set W_i referred to in Chapter 1 might consist of only two constants and a description of concrete in terms of those two should meet the criteria discussed in Chapter 1.

Some idea of the ambiguities that can arise if only one measurement is made can be obtained from Figure 3.7 which shows the flow curves of two Bingham materials labelled A and B, whose flow curves cross. If

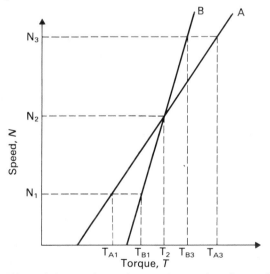

Figure 3.7 Effect of choice of speed upon the results of a single-point measurement on two Bingham materials.

a measurement is made at speed N_1 it will be concluded that material A is thinner than material B because the torque is less for A than it is for B, whereas if a measurement is made at speed N_3 the opposite conclusion will be reached, and at speed N_2 the materials will be judged as having the same flow characteristics. It is obvious that if a single measurement test operates at speed N_3 and then the material is used in a process whose effective speed is N_1, difficulties are likely to be experienced.

Thus, if only a single measurement is made, that is if the test is a single-point test, it is not possible to obtain the flow curve or to deduce the values of the two constants that measure yield value and plastic viscosity. In other words, it is not possible to get out of the experiment more information than was put in. All that can be done in this case is to quote the single value obtained or to treat the material as Newtonian, when it is not, that is, to give the value of the slope of the line drawn from the single experimental point to the origin, as shown in Figure 3.8. The reciprocal of this slope is a measure of what is called the **apparent viscosity**, that is, it is the viscosity of a Newtonian that would behave in the same way as the Bingham under consideration **at the particular speed** of the experiment. Clearly, apparent viscosity depends not only on the properties of the material being investigated but also on the speed at which the single measurement is taken. When apparent viscosity is used in ignorance the result can be costly chaos, but under some circumstances it can be a useful quantity and it will enter other considerations later on.

The fact that concrete has a yield value is important in other ways too, for example, in the way that it behaves when it is pumped, and that can be best understood by considering what happens when a Bingham material is investigated in the capillary-tube type of viscometer. It has already been seen that a Newtonian that gives a straight-line relationship in the coaxial-cylinders viscometer also gives a straight line of the same type in the capillary tube, so it seems reasonable to expect that a Bingham that gives a straight line with an intercept in the coaxial cylinders will also give a straight line with an intercept in the capillary tube. However, that is not so.

Material flowing in a capillary tube may be regarded as doing so in coaxial cylindrical layers, rather like the continuous extending of a telescope, and the shear stresses to be considered are those acting on the cylindrical surfaces. It can easily be shown that these stresses are not uniform across the section but have values proportional to the radius of the cylinder under consideration. It follows that the shear stress is zero at the centre of the tube where the radius is zero, and rises to its highest value at the inner surface of the tube where the

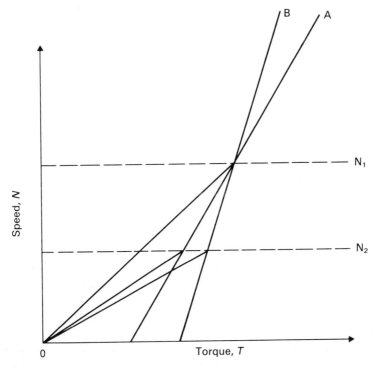

Figure 3.8 Apparent viscosities of two Bingham materials A and B: at speed N_1, η_{app} values are equal and are measured by the reciprocal of the slope of line OX; at any other speed, say N_2, η_{app} values are different from that at N_1 and different from each other.

radius is greatest. When a Bingham material is contained in a flow tube and the pressure to move it is gradually increased, it is at the wall of the tube that the applied stress first equals the yield stress, and at this moment it does not reach the yield stress anywhere else. The result is that shearing takes place in a thin layer close to the wall, and nowhere else, so the material moves forward as a solid plug, as illustrated in Figure 3.9. Not surprisingly, this phenomenon is known as **plug flow**. An everyday example of it is shown in the way that toothpaste flows out of the nozzle of its tube. As the pressure difference causing flow is increased, the yield value is reached at progressively smaller and smaller radii and the radius of the plug is steadily reduced, but, at least theoretically, it never actually becomes zero, so that material is flowing across the entire cross-section, until the pressure becomes infinite. The result is that the flow curve approaches, but theoretically never becomes, a straight line. This is one of the reasons why the capillary-tube viscometer is not a very good instrument for investigation of other

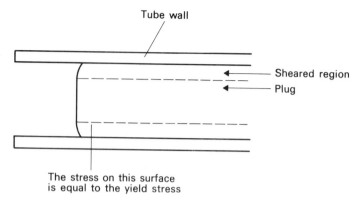

Figure 3.9 Plug flow.

than Newtonian materials, that is, because plug flow cannot be eliminated. Plug flow, or its rotational equivalent, does in fact occur in the coaxial-cylinders viscometer too, and causes a curvature of the bottom end of the flow curve. However, in this case it can be eliminated simply by not working at very low speeds and, above the relatively small region where plug flow occurs, the flow curve does not just approach a straight line, it is one.

Plug flow may be observed in some cases of the pumping of concrete, but there is an additional complicating factor. A material that fits the Bingham model is likely to be a suspension of solid particles in a fluid medium and it is possible for there to be some separation of the more fluid constituent towards the walls of an apparatus to form what are called slippage layers. Pumping of concrete will be discussed later but it may be noted now that plug flow and slippage layers are both important factors to be considered.

There is now no doubt whatsoever that the flow properties of fresh concrete approximate closely to the Bingham model and should therefore be assessed in terms of not fewer than two constants, the yield value and plastic viscosity. This statement is supported by an overwhelming amount of experimental evidence, some of which will be considered in the discussions that follow. It is clear that any attempt to measure workability and to understand the behaviour in practice will not stand much chance of success, unless it takes account of that fact.

4 Principles of measurement

The last chapter anticipated somewhat in stating as a simple fact that the flow properties of fresh concrete approximate to the Bingham model, without giving any details of the vast amount of experimental evidence that is now available in support. Of course, when the first attempt is made to investigate a material, the nature of its properties is not known or the investigation would be unnecessary. Nevertheless, it would be unusual to know nothing at all and in the case of fresh concrete it could be said that it would be unlikely to be a simple Newtonian, because it is a complex mixture of particles ranging in size from about a micron to say 20 mm, and some of the constituents are reacting chemically with each other. This should be sufficient to suggest that a single-point method that might be suitable for a Newtonian is not likely to be satisfactory but in any case, if it were a Newtonian, measurements at a series of speeds would be needed to establish that.

In addition, remembering that ordinary observation suggests that concrete possesses a yield value, the question of plug flow must be taken into account, and this is a good reason for rejecting apparatus of the flow-tube type. Another disadvantage of that type is that a continuous supply of concrete must be provided unless a complicated recirculation system is introduced. The most obvious candidate for initial investigations is some type of coaxial-cylinders viscometer. Although such viscometers had been used for work on cement pastes and mortars several decades ago, there does not seem to have been any attempt to use one for concrete before about 1970 when Tattersall[1] reported unsuccessful experiments in which he found that a failure plane developed in the concrete between the cylinders, and the measured torque was independent of rotation speed. After trying replacement of the inner cylinder by inner members of a wide range of shapes, without success, he abandoned this approach in favour of the use of a mixer method.

Concurrently, Uzomaka[2] had more success on concretes of much higher workability, of the type used for piling and diaphragm walling, and obtained evidence to suggest compliance with the Bingham model. The coaxial-cylinders apparatus, or a variant of it, was also used by Murata and Kikukawa[3], Morinaga[4], and Sakuta, Yamano, Kasami and Sakamoto[5] and more evidence to support the Bingham concept was obtained. The principle of measuring the torque required to initiate movement of a body immersed in the concrete had been used earlier by other workers such as l'Hermite[6], Ritchie[7] and Komloš[8], but they did not attempt to obtain flow curves or apply any theory.

All the work considered above, with the exception of Tattersall's which was a failure, and that in which only yield value was measured, was on concretes of fairly high workability. There is no report of the successful use of the coaxial-cylinders viscometer for concretes having workabilities in the range normal for structural concretes in the UK, and even the apparently successful work on the higher workabilities has given results that differ by orders of magnitude. Clearly, some searching questions need to be asked about this method.

Bloomer[9] has discussed in detail the requirements for the design of a coaxial-cylinders viscometer suitable for measurements on concrete, and points out that it is necessary to consider gap size, the ratio of the cylinder diameters, end effects, and possible slippage. The effect of this is to impose conditions on the dimensions of the apparatus and it is immediately apparent that none of the investigators mentioned above used viscometers that satisfied the criteria, especially so in the case of gap size. It has been stated by Van Wazer, Lyons, Kim and Colwell[10], and is generally accepted, that the difference between the radii of the inner and outer cylinders of apparatus to be used for a particulate suspension should be not less than ten times the size of the largest particle in suspension. This is a reasonable condition supported indirectly by work such as that of McGreary[11] who found in experiments on packing of particles in containers of various sizes, that the bulk density of the packed material was independent of the ratio of the size of container to size of particle provided that ratio was not less than ten.

Bloomer calculated that to fulfil all the criteria satisfactorily a viscometer for concrete would have to be very large and would require a sample of $2.6\,\mathrm{m}^3$, that is, half a ready-mixed truckload. Clearly this is quite impractical, but a reduction in size would be at the sacrifice of desired conditions and would therefore reflect on the reliability of results.

Because of these difficulties, and because results from experiments that have been done exhibit great variability, it is necessary to consider some alternative technique, and one possibility is to use measurements made during a mixing process.

Quite apart from any consideration such as that given above, the attraction of attempting to assess workability by making some measurement during the mixing process is obvious. Methods have been proposed by Purrington and Loring[12,13], Roberts[14], Rayburn[15], Hagy[16], Polatty[17], Simmonet[18], and Harrison[19,20]. However, nearly all these workers made measurements at one speed only. The only exceptions were Harrison, and Purrington and Loring, and in neither of these two cases was there any proposal to plot a flow curve or to obtain the constants needed to describe it. The first reported attempt to do so related to a laboratory method described by Tattersall[1] in 1970.

Tattersall's approach was also an empirical one and followed his experiments with a coaxial-cylinders viscometer in which a failure plane had developed in the concrete so that the value of the torque became independent of the speed of rotation of the cylinder. He reasoned that the problem could be avoided by the use of an apparatus in which the element on which the torque is measured is continuously

Figure 4.1 The original set-up using a Hobart foodmixer and a dynamometer wattmeter.

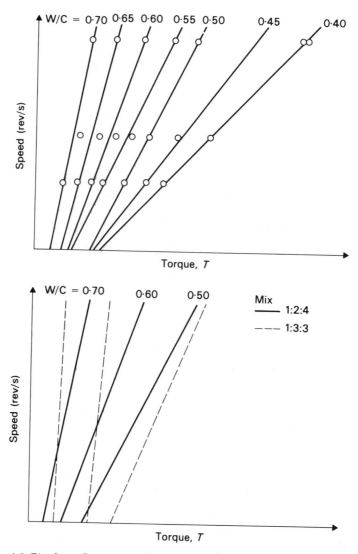

Figure 4.2 Bingham flow curves for concrete obtained in Hobart apparatus. (a) effect of changing water content for 1:2:4 mix; (b) effect of changing coarse/fine ratio.

presented with a new volume of concrete, and achieved this by using an ordinary Hobart food mixer equipped with a stirring hook that moved in planetary motion in a bowl whose volume was approximately 10 Litres. By choice of gears the hook could be given any one of three speeds, 1.6, 2.8, or 5.2 rev/s about the hook axis, with corresponding speeds of 0.48, 0.88 and 1.65 rev/s about the bowl axis. The apparatus is shown in Figure 4.1. The electrical power was measured by means of a dynamometer wattmeter when the bowl contained a standard quantity (25 kg) of concrete, and also when the bowl was empty. The difference between these two powers, P, was divided by speed, N, to give a value of torque, T, in arbitrary units, and T was then plotted against speed. In general, for concretes of practical interest, the relationship was linear or very nearly so; some typical results are shown in Figure 4.2.

This very simple result, of a linear relationship, is at first a very surprising one because it had been known at least since 1954[21-23] that the behaviour of cement/water pastes is very complicated and it might be expected that concrete, which may be regarded as a cement paste containing inert aggregate particles of a wide range of sizes, would be even more so. The fact that it is not can now be explained; briefly, it is because the shear rates imposed are very low.

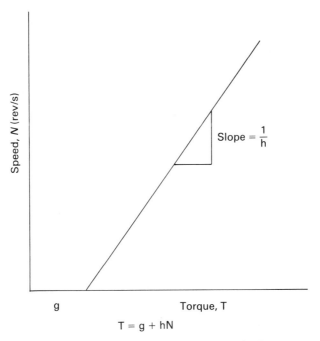

Figure 4.3 Relationship between torque and speed for fresh concrete showing that the material conforms to the Bingham model.
$T = g + hN$
Intercept g measures yield value.
Reciprocal slope h measures plastic viscosity.

The linear curves can be represented by the simple equation

$$T = g + h \cdot N \tag{4.1}$$

Where g is the intercept on the torque axis and h is the reciprocal of the slope of the line, as shown in Figure 4.3. It is immediately obvious that this is a form of the equation for the Bingham model, so that g is a measure of the yield value of the concrete and h is a measure of its plastic viscosity.

This preliminary work was followed by a careful and detailed investigation by Scullion[24] and further work by Dimond and Bloomer[9], backed up by trials in other laboratories. Useful results were obtained, some of which will be referred to later, but the most important conclusion was that while the principle of measuring the Bingham constants was sound, this particular apparatus had deficiencies such that it should be replaced by one specially designed for the job. That task was therefore undertaken and an apparatus as described in the following chapter was developed.

REFERENCES

1. Tattersall, G.H. (1971) Measurement of workability of concrete, Paper presented to a Conference of the East Midlands region of the Concrete Society, Nottingham.
2. Uzomaka, O.J. (1974) A concrete rheometer and its application to a rheological study of concrete mixes, *Rheologica Acta*, **13**, 12–21.
3. Murata, J. and Kikukawa, H. (1973) Studies on rheological analysis of fresh concrete, in *Fresh Concrete: Important Properties and their Measurement, Proceedings of a RILEM Seminar held 22–24 March 1973, Leeds*, Vol. 1. Leeds, The University, pp.1.2-1–1.2-33.
4. Morinaga, S. (1973) Pumpability of concrete and pumping pressure in pipe lines, *Fresh Concrete: Important Properties and their Measurement, Proceedings of a RILEM Seminar held 22–24 March 1973, Leeds*, Vol. 3, pp. 7.3-1–7.3-39.
5. Sakuta, M., Yamano, S., Kasami, H. and Sakamoto, A. (1979) Pumpability and rheological properties of fresh concrete, in *Proceedings of Conference on Quality Control of Concrete Structures, 17–21 June 1979, Stockholm*, Vol. 2. Stockholm, Swedish Cement & Concrete Research Institute, 125–32.
6. L'Hermite, R. (1949) Recent Research on concrete, *Annales des Travaux Publics de Belgique*, **50**(5), 481–512, (Library Translation Cj. 24, London, Cement & Concrete Association, 1951).
7. Ritchie, A.G.B. (1967) The Vane method of measuring the mobility of fresh concrete. Paper presented to Conference of the British Society of Rheology, Sheffield.
8. Komloš, K. (1966) Penetrator for testing of concrete mix workability, *Stavebnícky Časopis*, **14**(9), 571–2.
9. Bloomer, S.J. (1979) Further development of the two-point test for the measurement of the workability of concrete. PhD Thesis, University of Sheffield.

10. Van Wazer, J.R., Lyons, J.W., Kim, K.Y. and Colwell, R.E. (1963) *Viscosity and Flow Measurement*, New York, Interscience.
11. McGreary, R.K. (1961) Mechanical packing of spherical particles, *Journal of the American Chemical Society*, **44**, 513.
12. Purrington, W.F. and Loring, H.C. (1928) The determination of the workability of concrete, *Proceedings of the American Society for Testing Materials*, **28**, Part II, 499–504.
13. Purrington, W.F. and Loring, H.C. (1930) Further studies on the workability of concrete, *Proceedings of the American Society for Testing Materials*, **30**, Part II, 654–73.
14. Roberts, E.D. (1931) Determining characteristics of concrete in the mixer drum, *Journal of the American Concrete Institute, Proceedings*, **28** (9), 59–72.
15. Rayburn, E.B. (1934) Consistency indicator for a ready mixed concrete plant, *Journal of the American Concrete Institute, Proceedings*, **31**(2), 105–12.
16. Hagy, E.A. (1937) The apparatus for testing concrete mixtures for consistency, US Patent 2089604, Washington DC, US Patent and Trade Mark Office, 9 pp.
17. Polatty, J.M. (1949) New type of consistency meter tested at Allatoona Dam, *Journal of the American Concrete Institute*, **21**(2), 129–36.
18. Simmonet, J. Self operating device for the manufacture of constant workability of concrete, in *Fresh Concrete: Important Properties and their Measurement, Proceedings of a RILEM Seminar held 22–24 March 1973*, Leeds, Vol. 1, pp. 2.9-1–2.9-50.
19. Harrison, O.J. (1964) An investigation into the relationship between concrete workability and the pressures in the hydraulic systems of truck mixers. Technical Report No.27, Feltham, Middx, Ready Mixed Concrete (UK) Ltd.
20. Ready Mixed Concrete Ltd. Improvements in or relating to apparatus for gauging the consistency of wet concrete. British Patent 1223558.
21. Tattersall, G.H. (1954) The rheology of cement pastes, fresh mortars and concretes. Thesis submitted to the University of London for the degree of MSc, 174 pp.
22. Tattersall, G.H. (1965) The rheology of Portland cement paste. *British Journal of Applied Physics*, **6**(5), 165–7.
23. Tattersall, G.H. (1955) Structural breakdown of cement paste at constant rate of shear, *Nature*, **175**(4447), 166.
24. Scullion, T. (1975) The measurement of the workability of fresh concrete, MA Thesis, University of Sheffield.

5 The two-point workability test

5.1 PRINCIPLES

The objective in developing the new apparatus was, of course, to exploit the principle of expressing the flow properties of concrete in terms of two constants, which had been established as useful and promising by the work already completed, but to avoid the disadvantages and defects of the earlier apparatus. This would be done by obtaining a flow curve from measurements of the torque required to rotate a suitable impeller immersed in the concrete, at several different speeds.

If it is known with confidence that the concrete under test behaves according to the Bingham model, and if experimental error is ignored, it is necessary only to measure torque at two speeds, and that is why the test is called the two-point test. In practice, the use of a range of speeds not only permits a proper check to be made on the linearity or otherwise of the relationship, but it also, if the relationship is linear, reduces the experimental error in the determination of the two constants g (the intercept) and h (the reciprocal slope).

Since the apparatus was intended for plant and site use, there were several other requirements.

(a) The quantity of concrete needed for test should be large enough to be representative but small enough to be handled easily, say one bucketful.

(b) The apparatus should be simple and robust.

(c) It should be as cheap as possible consistent with giving a satisfactory performance.

(d) It should be capable of being fabricated on a one-off basis so that its future development would not be retarded by a need for expensive tooling-up.

(e) Torque measurement should be as simple as possible but must not be affected by contamination by dust.

(f) Because it seemed unlikely that one particular arrangement would be suitable for the whole range of workabilities, the necessity for theoretical treatment, so that results from different arrangements could be related to each other, must be borne in mind.

(g) Future possibilities of linking up the measuring system to recorders or computers should also be considered.

5.2 DEVELOPMENT

The apparatus was first developed for use with medium- to high-workability concretes.

The bowl to contain the sample under test is in the form of a cylinder of 254 mm diameter and 305 mm high (10 in × 12 in) so that it can be fabricated from flat sheet, and requires a sample of about a bucketful. The impeller must move the concrete without causing or exaggerating segregation and/or bleeding, and after trials of various designs, the interrupted helix form shown in Figure 5.1 was finally selected. This shape has the advantages that it is made from flat blades fixed in a helical thread cut in the central shaft and that it permits concrete to fall back through the gaps between the blades. This latter feature is particularly important; if a complete helix were to be used the material that is moved upwards would not immediately be replaced by concrete from the sides. The sense of rotation of the impeller is such (anti-clockwise) as to raise the concrete.

An arrangement that gave the required torque and speeds was obtained by the use of a $\frac{1}{2}$ hp single-phase electric motor driving through a hydraulic transmission (Carter Gear F10) and a 4.75 : 1 right-angled reduction gear, which provides a torque range of 0–16 Nm and an infinitely variable speed from −3.15 through zero to +3.15 rev/s.

Another important reason for the selection of the Carter gear as the driving unit arose from the requirements for torque measurement, because the torque developed is quite accurately proportional to the pressure of the oil in the hydraulic transmission. Most methods of torque measurement are either cumbersome, or expensive, or suffer from the disadvantage of susceptibility to interference by dirt, whereas use of this property of the hydraulic unit provides a method that is exempt from all these disadvantages. All that is necessary is to connect a simple pressure gauge in the hydraulic line to provide a system that is robust enough for site use and is maintenance-free. Alternatively, or in addition, a simple pressure transducer may be used for conversion of the pressure reading to a digital read-out or for coupling to a chart recorder or, if desired, to a computer.

One of the many alternatives considered was the use of strain

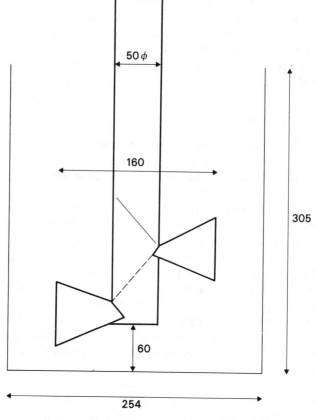

Figure 5.1 Interrupted-helix impeller.

gauges fixed to the drive shaft but this solution was rejected on the grounds that not only is it more expensive but it requires the provision of sealing glands to protect the necessary slip rings, and would certainly need at least occasional skilled maintenance. Contamination by the dirt that is inevitably present under the conditions of use intended also seemed to be very likely and this would result in erratic and unreliable results. Cabrera and Hopkins[1] have subsequently used such a system successfully and report that no trouble has been experienced, but it may be noted that this is under laboratory conditions only.

Wallevik and Gjørv had tried the strain gauge and slip rings method in 1983[2] but found interference caused by mechanical vibrations, and they quote Rohrbach[3] as warning that stray signals are created in the interface between the brushes and the slip rings. They continue to use the hydraulic method with a pressure transducer.

Szabowski and Jastrzebski[4] recently tried another electrical method of measuring the torque on the impeller shaft and managed to avoid the difficulties associated with slip rings by using some non-contacting method of transmitting the signals to a read-out. They also altered the method of speed control to one involving mechanical gears and electrical control of the d.c. motor that provided the power. No arguments are given to show that these arrangements have any advantages over those of the original apparatus on which the essentials of their design are based. They give results from only one mix and those do not differ in any important particular from those of previous workers.

The original pressure-gauge system with the hydraulic unit has amply justified the faith placed in it by successful use on many sites and in many ready-mixed concrete plants.

5.3 THE APPARATUS

The final arrangement of the prototype apparatus is shown in the diagram of Figure 5.2 and the photograph of Figure 5.3. The bowl containing the concrete is supported on a suitable arm that can be raised or lowered by means of a rack and pinion. The impeller, in the form of the interrupted helix described above, is immersed in the concrete and is rotated about its own axis, coaxial with the cylinder. The pressure gauge ($0-7\,\mathrm{N/m^2}$, $0-1000\,\mathrm{lb/in^2}$), backed by a thick rubber mat to reduce the effects of vibration, is connected to the hydraulic gear box by a flexible hose and, to cut down on oscillations, an adjustable valve is provided in the hydraulic line. All the parts are mounted on a simple frame, fabricated from steel angle section, which is provided with adjustable feet for levelling and castors for ease of movement.

The speed control knob provided on the hydraulic unit is already calibrated but it was found that at a given speed setting the actual speed decreases slightly as torque increases. The effect is so slight that it is probably not important for site control and it could in any case be allowed for in the calibration constant of the machine, but the simplest and most satisfactory course is actually to measure the speed in every test and this is done by means of a reflecting tachometer.

The apparatus in this form has been found to be fully satisfactory for medium- to high-workability concretes (so is known as the MH form), and it was immediately used in dealing with practical industrial problems. However, it was found to be unsuitable for low-workability

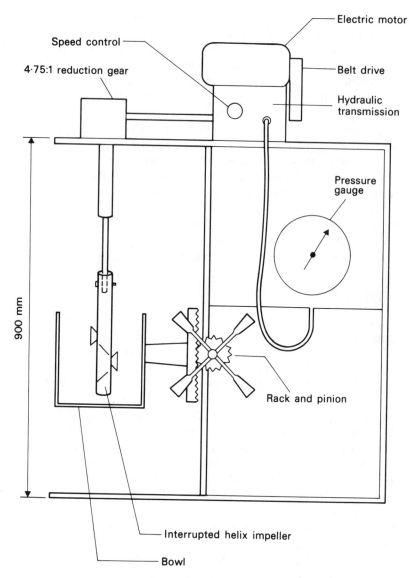

Figure 5.2 Two-point workability apparatus.

concretes, of slump lower than about 50 mm or so, because the uniaxially rotating impeller tended to move the concrete to the side of the bowl and then rotate in the resulting hole, that is, the concrete failed to fall back into the centre. Evidently, some modification was needed.

Figure 5.3 Two-point workability apparatus.

5.4 MODIFICATION FOR LOW-WORKABILITY CONCRETES

It was decided that for low-workability concretes it would be necessary to introduce planetary motion of the impeller, but with the constraint that any change should be as simple as possible and that two other conditions should be met. The first of these was that the planetary gearing ratio should not be a whole number, so that the impeller does not return to the same position after each revolution of the central shaft and is thus more likely to dislodge any concrete that otherwise might remain at the side of the bowl. The second was that the degree of offset of the impeller shaft from the centre line of the bowl, and the impeller-bowl clearance, should be such that the amount of concrete required for a test did not become excessive while the occurrence of aggregate trapping, between the impeller and the bowl, must be acceptably low.

These considerations resulted in the design of a planetary gear box consisting simply of a steel block which clamps on the drive shaft of the workability apparatus and carries in suitable bearings a shaft equipped at its upper end with a gear that engages with a larger gear bolted to the bearing housing. This is shown in Figure 5.4. The gear ratio is 2.25, thus satisfying the requirement that it should not be a whole number, and the offset distance is 50 mm so that a bowl of larger diameter, 356 mm (14 in) must be used. The size of concrete sample needed is therefore increased to about $1\frac{1}{2}$ bucketfuls but the loaded

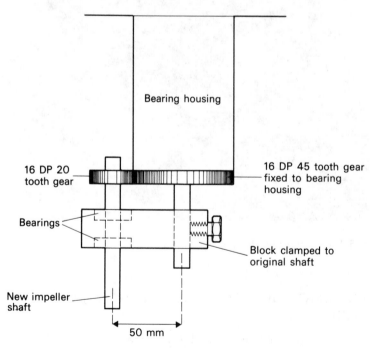

Figure 5.4 Planetary gear.

bowl can still be handled by one man. It was also found necessary to change the impeller shape, and after trials the H shape shown in Figure 5.5 was selected. This impeller fits on to the planetary gear shaft and is held by a pin, in the same way as the interrupted helix impeller is fixed on to the drive shaft in the MH form of the apparatus. This new form is known as the LM form, because it is suitable for low to medium workabilities and is shown in Figure 5.6.

Because the planetary gearing is also in effect a step-up gear, the 4.75 ratio reduction gear of the MH form must be changed to a 20:1 gear. All these changes can be made in about five minutes and that time could be reduced to a minute or so by the introduction of a further gearbox, if the extra cost can be justified.

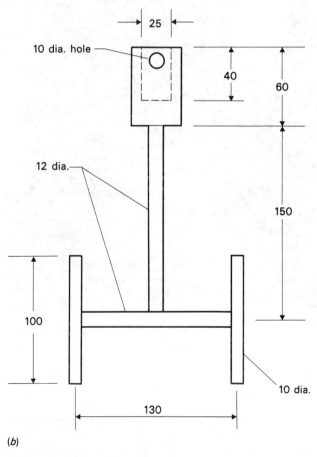

(b)

Figure 5.5 H impeller (dimensions in mm).

Figure 5.6 Two-point workability apparatus with planetary motion of H impeller.

5.5 USE OF THE APPARATUS

Before use the reduction gearbox and the hydraulic unit must of course be filled with the appropriate oil and the latter must be bled to remove entrapped air. If starting from cold, the apparatus is allowed to warm up by rotating the impeller, before testing, for about 30 min. The speed recommended for this operation was 0.7 rev/s (speed setting 2) but Wallevik and Gjørv[2] state that at this speed the idling pressure can still be changing even after 80 min, so they recommend the considerably higher speed of 3 rev/s (speed setting 8). After the warm-up period, the procedure, in the MH mode, is as follows.

(a)   Raise bowl to working position, which is such that the clearance between the bottom of the impeller shaft and the bottom of the bowl is 60 mm.

(b)   With the impeller rotating at about 0.7 rev/s fill the bowl gradually to approximately 75 mm from the rim with concrete.

(c)   Increase speed setting to approximately 1.3 rev/s and allow time for pressure to stabilise (i.e. at speed setting 4)

(d)   Read speed by tachometer.

(e)   Read pressure gauge. Large oscillations due to trapping of aggregate should be ignored and the average position of the needle for the small oscillations should be recorded (see later).

(f)   Repeat (d) and (e) at speeds of approximately 1.2, 1.0, 0.9, 0.7, 0.5, and 0.3 rev/s, (i.e. speed settings $3\frac{1}{2}$ to 1 in steps of $\frac{1}{2}$).

(g)   Remove the bowl containing the concrete and record the idling pressure at each of the speeds used in the measurements on the concrete.

The decision about how many speeds to use should be made by balancing any constraints imposed by the time available against the advantage of lower experimental error associated with a greater number.

The procedure for the LM mode (i.e. planetary motion) is essentially the same as for the MH mode but the clearance between impeller and bowl is 90 mm and the speeds used should cover the range from 1.9 to 0.7 rev/s (i.e. speed settings 6 to 2).

5.5.1 Calculation of results

Because of the accumulated evidence that the flow properties of fresh concrete conform to the Bingham model, in nearly all practical cases, there is no need to check by drawing that the flow curve is linear. All that is necessary is to obtain the best straight-line relationship between torque and speed so that from it the values of g, the intercept on the torque axis, and h, the reciprocal of the slope of the line, may be found. The best line is taken as being that for which the sum of the squares of the deviations of the dependent variable (in this case the torque) is a minimum, that is, the one that is commonly known as the **least squares line**, and it is easily calculated using an inexpensive calculator. The worked example given in Table 5.1 should make it clear.

Table 5.1 Calculation of g and h from experimental data obtained using the two-point apparatus (calibration constant 0.019 Nm/pressure unit)

| Speed (rev/s) | Pressure in arbitrary units | | | Torque (Nm) |
	Total	Idling	Net	
1.39	465	97	368	6.99
1.28	455	93	362	6.88
1.11	440	87	353	6.71
0.94	410	83	327	6.21
0.75	395	77	318	6.04
0.56	375	73	302	5.74
0.38	360	67	293	5.57

Enter pairs of values of speed and torque into the calculator, putting in speed as x and torque as y. Read off correlation coefficient, intercept and slope. Note that by inputting speed as x, h is found in the calculator as slope – not reciprical slope.

Result   Correlation coefficient $r = 0.990$
　　　　　Intercept g　　　　　　4.93 ± 0.2
　　　　　Slope h　　　　　　　　1.49 ± 0.2

(see later for calculation of experimental error).

The calculation is shown as above for clarity. It can, of course, be carried out more quickly by applying the calibration constant at the end instead of throughout, i.e. by calculating the best line between pressure and speed (instead of between torque and speed), and then multiplying the intercept and slope by the calibration constant to give g and h.

5.5.2 Calibration of the apparatus

The calibration constant for conversion of pressure readings to torque will normally be given with the apparatus but it can easily be determined using any method by which known torques can be applied.

It is not necessary to measure idling pressures every time a measurement of workability is carried out. They should be measured at the start of a day's work to check that the warming-up period has been adequate, and then a few more times during the day when time permits.

In experiments at temperatures from 5 °C to 25 °C Wallevik and Gjørv[2] showed that the calibration constant does depend on temperature and it can be calculated from their results that the temperature coefficient is about 1.3% per °C. This effect will not be important on normal sites or in normal plants over a short period, but it will be if even UK extremes of temperature are encountered, and it should be considered.

Table 5.2 Experimental error

Correlation coefficient	Percentage error in value of h
0.999	4.0
0.998	5.7
0.997	7.0
0.996	8.1
0.995	9.1
0.994	9.9
0.993	10.7
0.992	11.5
0.990	12.9
0.985	15.8
0.980	18.3
0.975	20.6
0.970	22.6

5.6 EXPERIMENTAL ERROR

Because the values of g and h are derived from a linear relationship obtained as the best equation for a number of experimental points, it is possible to assign an estimate of experimental error to the result of every separate determination.[5] To do this, use is made of the value of the correlation coefficient which is obtained from the calculator as a result of the same operation that gives the values of g and h.

If the number of experimental points is seven, the percentage error on h is obtained from Table 5.2 and the absolute error on h can easily be calculated. This figure also gives the absolute error on g.

Thus, in the worked example of Table 5.1, seven experimental points gave $g = 4.93$, $h = 1.49$, $r = 0.990$. From Table 5.2, the percentage error on h is 12.9 so the absolute error is 1.49×0.129 which is 0.19 and that is also the absolute error on g. The results of the test may therefore be written as $g = 4.93 \pm 0.2$, $h = 1.49 \pm 0.2$. Even with the same correlation coefficient, if the number of experimental points is less, then the estimated experimental error is greater. For fewer than seven points the method of obtaining the errors is exactly the same but then

Table 5.3 Correction factors

Number of points	Correction factor
6	1.05
5	1.16
4	1.45

the results must be multiplied by the appropriate correction factor obtained from Table 5.3.

In practice, it should be possible to obtain correlation coefficients of about 0.99 with an associated error (for seven points) of about 12%.

In a series of 16 test programmes, including altogether 1065 test results, Wallevik and Gjørv[2] obtained average correlation coefficients from 0.984 to 0.995 (with standard deviations of 0.0059 and 0.0181 respectively) and from this and other evidence they concluded that reproducibility was very good.

In has been shown[5] in properly designed factorial experiments that results obtained do not depend on the operator, even when one is inexperienced, or in other words, the test is not operator-sensitive. This result has in general been confirmed subsequently by Wimpenny and Ellis[6] although they do find that it is not true for an operator who has had literally no previous experience at all. The practical answer to this is simply to ensure that a new operator carries out about half a dozen practice runs.

5.7 OSCILLATION OF TORQUE

It has already been stated that a valve is provided in the hydraulic line of the apparatus, to cut down on oscillations of the pressure gauge. This valve must be adjusted to an optimum setting by simple trial and error and must not be closed to an extent such that over-damping of the needle occurs so that its response is sluggish, but the adjustment is fairly easy to make. Nevertheless, oscillations cannot be eliminated completely and the consequence is that an operator new to the machine needs a short learning time, to acquire the necessary experience.

The oscillations have two components, which are usually easily distinguishable; there is an oscillation of relatively low amplitude (typically of the order of a few per cent of the mean) upon which are superimposed larger kicks caused by trapping of the aggregate. It is the mean of the former that should be taken as the pressure reading and it has been convincingly demonstrated on many occasions that an unskilled operator can learn in a very short time (e.g. $\frac{1}{4}$ hour) how to do this simply from visual observation of the pressure gauge.

Of course, the process can be made even easier by incorporating a pressure transducer in the hydraulic line so that a recording of pressure as a function of time may be produced. This was first done by Gjørv and co-workers[7].

Cabrera and Hopkins[1], whose apparatus was equipped with a strain-gauge torque-measuring system, as mentioned earlier, also obtained torque/time traces but the chart speed they used was so low that individual oscillations could not be distinguished and the trace degenerated into a wide band from which little information could be obtained. Cabrera and Hopkins took the centre line of the band (which they wrongly described as the median value) as giving the torque value to be used in calculations and in doing so they proposed a procedure different from that used in all other work with this method.

A useful and thorough investigation was carried out by Wimpenny and Ellis[6] who used the pressure-transducer method and worked with the apparatus in its LM, or planetary, mode because it is for this form that the oscillations are worst. Typical traces obtained by them are shown in Figure 5.6 for a low-workability mix (c. 50 mm slump) and a higher-workability mix (c. 100 mm slump). The kicks are worse in the former case than in the latter but in both it is easy to distinguish them from the underlying small oscillation.

Wimpenny and Ellis made a detailed examination of many traces and, because their equipment was connected to a computer, they were able to calculate in each case three different quantities that might be proposed as characteristic measures of the trace. These were:

B. the mean of all readings taken at the rate of 4 per sec,

C, the mean of all readings less the kicks, and

E, the median (i.e. middle value) of all readings,

and they compared these with a value obtained subjectively as the representative value D of the trace by a completely independent observer. Figure 5.7 demonstrates that C, D and E were in close agreement, that is, subjective assessment from the trace is highly satisfactory. This subjective assessment from the trace also agreed well with direct pressure-gauge readings for an experienced operator and slightly less so for one who had had no previous experience at all, and for both operators the values of g and h calculated from their observations were within ±5% of the mean.

This confirms that the practice of reading from the pressure gauge is satisfactory, once a little experience has been gained, but there is an advantage to be gained, in ease of operation, by using a simple recorder in conjunction with a pressure transducer, particularly for low-workability mixes. In practice, oscillations are somewhat more troublesome for the planetary mode than for the uniaxial, for lower workabilities, and for larger maximum-size aggregates.

The apparatus described was developed for use with mixes having a maximum aggregate size not exceeding 20 mm but there is no reason why the bowl and impeller should not be scaled up for larger sizes. So far, only a very limited trial has been carried out, with a scaling-up factor of 1.4, but it did indicate that there should be no practical difficulty.

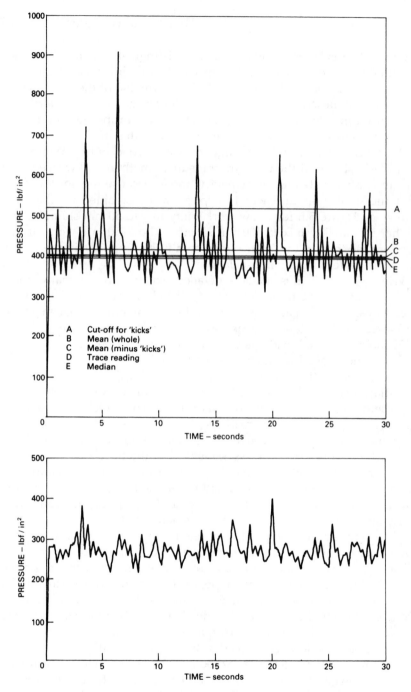

Figure 5.7 Typical total pressure traces using the LM apparatus (i.e. planetary motion of H impeller at 0.95 rev/s) (a) for a low-workability mix; (b) for a high-workability mix.
($1\,\text{lbf}/\text{in}^2 = 6.9\,\text{kPa}$) (*Wimpenny and Ellis*)

5.8 THEORETICAL CONSIDERATIONS

It has been shown that the flow properties of concrete can be represented by the equation

$$T = g + h \cdot N \qquad (5.1)$$

and it has been stated that the intercept g is a measure of the yield value while the reciprocal slope h is a measure of plastic viscosity, these being the two quantities that are sufficient to describe a material whose behaviour fits the Bingham model. So far, however, no theoretical justification has been given, and also, it should be clear that although g and h do provide measures of yield value and plastic viscosity respectively, their actual numerical values depend also on characterisitics of the apparatus, specifically, on the design and dimensions of the impeller and bowl. If the requirements listed earlier, in Chapter 1, are to be fully satisfied, some way must be found to eliminate this dependency.

The main difficulty in applying a theoretical treatment to the mixing process is that the rates of shear in the bowl vary very much from point to point. However, progress may be made by adopting the assumption of Metzner and Otto[8] that there is some effective average shear rate that is simply proportional to the speed of the impeller, that is

$$\dot{\gamma}_{\text{avg}} = K \cdot N \qquad (5.2)$$

It can then be shown[5] that the equation of a material conforming to the Bingham model is

$$T = (G/K)\,\tau_0 + G \cdot N \cdot \mu \qquad (5.3)$$

Where τ_0 and μ are respectively the yield value and plastic viscosity in fundamental units (i.e. independent of the details of the apparatus used to measure them), K is the constant in equation 5.2 and G is a constant. The values of K and G may be found by a rather complicated calibration procedure that involves making measurements on other materials whose rheological behaviour is already well-defined, and in particular, it means using a series of Newtonian liquids, and other liquids whose apparent viscosities are shear-rate-dependent in a known way.

It follows from equations 5.1 and 5.3 that

$$\tau_0 = (K/G) \cdot g \qquad (5.4)$$

$$\mu = (1/G) \cdot h \qquad (5.5)$$

So both g and h may be converted to fundamental units if desired. For practical use in industry this is not necessary, and the effort required to carry out the calibration would not be justified. The practical answer is to standardize the shape and dimensions of the bowl and impeller so that different pieces of apparatus give results in good enough agreement, and that is easy to do.

However, another advantage of the theoretical treatment combined with detailed calibration is that it makes it possible to relate to each other the results from different forms of apparatus. The relationship that is of particular interest, and of practical importance, is that between the results from the uniaxial form that is used for medium to high workabilities, and the planetary form that is used for low to medium workabilities. Figure 5.8, from work by Dimond, shows experimental points obtained on concretes of a wide range of workabilities, together with the theoretical line calculated after detailed calibration of each of the two forms of apparatus, and it can be seen that the agreement is fairly good. Dimond himself expresses some reservations about the validity of the theoretical line because of some difficulties concerning the calibration fluids, and Saeed[9] has found similar problems, so further work needs to be done. Nevertheless, results from the two forms of apparatus can be related and in practice they may be compared directly. The important practical consequence is that together they can cover almost the whole range of workabilities from low workability to flowing concrete and, moreover, on the same scale of measurement, an achievement that cannot be matched by any other method of workability measurement yet proposed.

In Dimond's work, the impeller used in the LM, or planetary, apparatus was in the shape of a propeller, and Figure 5.8 shows that the values of both g and h are about three times those obtained in the MH, or uniaxial, apparatus. The impeller now used in the LM mode is the H type described above; the relationship between the present two forms of apparatus is that the LM apparatus with the H impeller gives values of g about the same as those in the uniaxial MH apparatus, while the values of h are about one third higher.

5.9 RESULTS

Results from the two-point apparatus will be used extensively in the discussions that follow but, in general, actual flow curves will not be shown because, as pointed out earlier, it is not usually necessary to draw them if the correlation coefficient is satisfactory. It is therefore appropriate to give an example here.

The flow curves shown in Figure 5.9 were obtained by a previously inexperienced operator in his first attempts at using the apparatus[10].

The apparatus has now been used in many investigations by many independent workers and literally thousands of flow curves have been

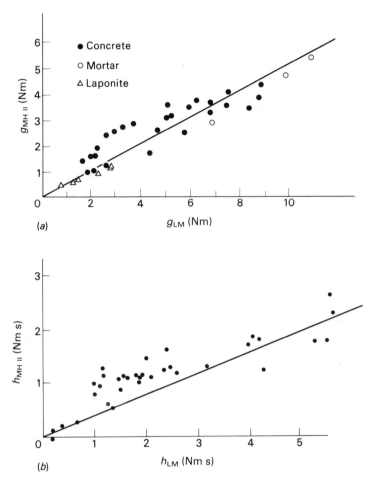

Figure 5.8 Correlation between (a) g values and (b) h values obtained in the MH (uniaxial) and the LM (planetary) forms of the two-point workability apparatus. (Dimond)
Note that the lines drawn are the theoretical relationships.

obtained. It can be regarded as being established beyond doubt that the flow properties of fresh concrete approximate closely to the Bingham model, that is, the relationship between torque and speed is a simple straight line with an intercept on the torque axis as shown in Figure 4.3 of the last chapter. This line is described by the simple relationship

$$T = g + hN$$

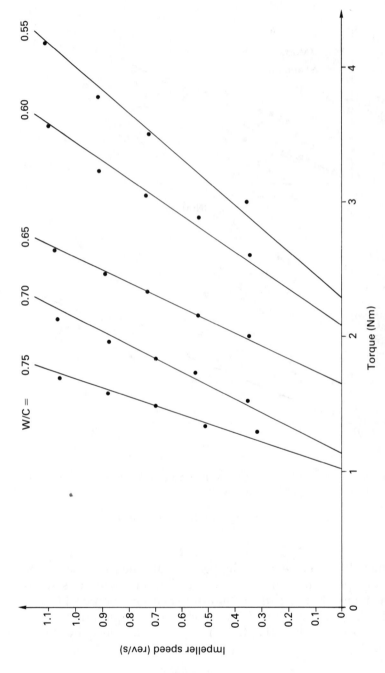

Figure 5.9 Effect of increasing water/cement ratio on the workability. Results obtained by inexperienced operator.

Where T is the torque at a speed N, g is the intercept on the torque axis and h is the reciprocal of the slope of the line. The quantity g is a measure of the yield value of the concrete and the quantity h is a measure of its plastic viscosity, both of which are fundamental properties of the material.

5.10 REFERENCES

1. Cabrera, J.G. and Hopkins, C.G. (1984) A modification of the Tattersall two-point test apparatus for measuring concrete workability, *Magazine of Concrete Research*, **36** (129), 237–40.
2. Wallevik, O.H. and Gjørv, O.E. (1990) Modification of the two-point workability apparatus, *Magazine of Concrete Research*, **42** (152), 135–42.
3. Rohrbach, C. (1967) *Handbuch für elektrisches Messen mechanischer Grossen*, VDI-Verlag GmbH, Dusseldorf, 336–42.
4. Szwabowski, J. and Jastrzebski, Z. (1988) Reometr Rod-l Pomiarow Reologicznych Zapraw I Mieszanek Betonowych, *Zeszyty Naukowe Politechniki Swletokrzyskiej*, **26**, 59–67.
5. Tattersall, G.H. and Banfill, P.F.G. (1983) *Rheology of Fresh Concrete*, London, Pitman.
6. Wimpenny, D.E. and Ellis, C. (1987) Oil pressure measurement in the two-point workability apparatus, *Magazine of Concrete Research*, **39** (140), 169–73.
7. Gjørv, O.E. (Jan. 1983) Private communication to G.H. Tattersall.
8. Metzner, A.B. and Otto, R.E. (1957) Agitation of non-Newtonian fluids, *Journal of the American Institution of Chemical Engineers*, **3**, 3–10.
9. Saeed, Ahmad (1982) Workability measurement with particular reference to the control of concrete production, PhD Thesis, University of Sheffield.
10. Al-Shakhshir, A.T. (1988) Workability of plasticized concrete. Dissertation submitted in part fulfilment of requirements for the Degree of MSc (Eng), University of Sheffield.

6 Workability expressed in terms of two constants

It has now been shown that the flow properties of fresh concrete approximate closely to the Bingham model, over the range of shear rates important in practice, and that, in terms of the measurements made in the two-point workability apparatus, this is expressed by the equation

$$T = g + hN \tag{6.1}$$

Thus, the relationship between torque and speed is a simple straight line that has an intercept g on the torque axis and a slope of $1/h$, where g is a measure of yield value and h of plastic viscosity. This result has been amply confirmed and there is now no doubt about it, so proper consideration should be given to its significance in terms of practical application.

The fact that two figures (not just one) are needed to describe the flow properties of fresh concrete requires a new approach to thinking about workability. Equation 6.1 is one way of stating the fact that the total physical effort to work fresh concrete is compounded of two terms, of which the first is of fixed value (for a particular mix) but the second increases as the speed of movement increases. The first term represents the effort to start the concrete moving at all and is quantified by the yield value, measured by g, while the second represents the extra effort to get it moving at a reasonable speed and is quantified by the plastic viscosity, measured by h, multiplied by the speed of movement.

This type of relationship is quite common in other everyday circumstances. For example, the annual cost of running a car (if depreciation is neglected) is the sum of fixed costs (such as tax, insurance, garaging) which depend on the type of car, and other costs (petrol, oil, tyres, clutch and brake wear) which depend on the type of car but also are roughly proportional to the distance travelled. Nobody would consider that a comparison of costs for different types of car would be adequate if it ignored one or other of these two constituents, or if it considered the total costs at some arbitrarily chosen mileage different from the mileage likely to apply in practice.

It is just as unacceptable to ignore one or other of the two terms that are important for workability assessment or to try to use results from some single measurement test that operates at some arbitrary (and unknown) shear rate different from the one that will apply in practice.

The argument that, for a Bingham material, measurements must be made at not fewer than two different shear rates, or speeds, has already been introduced in Chapter 3, but that argument is so important to the practical use of concrete that it is worth repeating and developing further.

Equation 6.1, which is the equation of the flow curve, contains the two quantities g and h, and these are the two quantities it is necessary to find in order to describe the flow properties of the concrete. Looking at the matter algebraically, it is an elementary fact that a solution for two unknowns can be obtained only if at least two simultaneous equations are set up. In this case, that may be done by measuring the torques T_1 and T_2 at the two different speeds N_1 and N_2 to provide the two equations

$$T_1 = g + hN_1$$
$$T_2 = g + hN_2 \qquad (6.2)$$

which may then be solved for g and h. It is clear that if only one measurement is made, so that only one equation can be set up, no solution is possible, and all that can be done is to quote the torque at the particular speed that was arbitrarily chosen for making the measurement. Such a result is useless as an indicator of the torque that would correspond to some other speed.

Figure 6.1 puts the same thing in geometrical form as already given

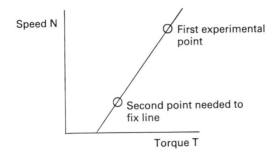

Figure 6.1 Flow curve of concrete.

in Chapter 3 and Figure 3.6. If only one experimental point is available the flow curve cannot be determined because there is an infinite number of lines that will pass through a single point; a second experimental point is essential to discover which one of them is the correct one.

It should not be thought that this argument applies only to the special case of a test in which torque is measured as a function of speed. Any test on the flow properties of concrete must necessarily work at some effective speed, even if that speed is unknown, and any test in which only one measurement is made is operating at only one effective speed. Thus, any test in which only one measurement is made, that is, a single-point test, is inherently incapable of providing sufficient information to describe the flow properties of concrete.

There is no way of dodging this; it does not matter how complicated the test or how sophisticated the treatment of results, it is not possible to get out of an experiment more information than is put in. The requirement for two constants is not just a matter of academic nicety or pedantry, it is a matter of great practical importance for the proper control and use of concrete. Anyone who thinks it can be dismissed as an ivory-tower concept will continue to be unable to measure workability, and will have to accept the possible occurrence of costly and time-consuming mistakes in practice.

The point is illustrated further in Figure 6.2 which is only another way of presenting Figure 3.7 of Chapter 3. Consider the two concretes A and B whose flow curves are as shown and cross at a point that is in the region of shear rates that correspond to practice. Suppose a single-point test is used to compare their workabilities. Whatever the type of test chosen it will measure either the amount or rate of defor-

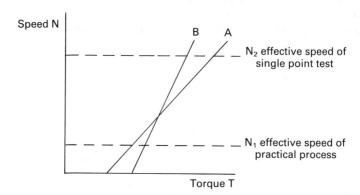

Figure 6.2 Misleading result obtained if effective speed of single-point test differs from that of process.

mation under given stress conditions, or the stress for a given deformation, so it is reasonable to associate it with some effective shear
rate that corresponds to some speed, say N_2, in the two-point test.
Since in this case the speed N_2 is above the cross-over point the test
will characterize concrete B as having a higher workability than concrete A, because the torque (or stress) required at this speed is lower
for the former than for the latter. Now suppose those two concretes are
used in a process whose effective shear rate corresponds to speed N_1,
a condition under which concrete B has a **lower** workability than
concrete A, so the result of the test is not just inadequate, it is seriously
misleading.

This illustration should be sufficient to dispose of the view, that has
sometimes been expressed, that this is all very well for the laboratory
but what is needed in practice on site or at the plant is a single figure
for workability. The requirement for two figures is inescapable; the
quantities yield value and plastic viscosity, measured by g and h, are
dimensionally different and it is no more sensible to ask for a single
constant to represent both of them than it is to ask for a single constant to represent both Young's modulus and Poisson's ratio of the
hardened concrete.

Note that if a single-point test has an effective shear rate corresponding to the speed where the two flow curves cross, it will indicate that
the workabilities of the two concretes are the same. It is well known
in practice that any one of the standard tests may classify as being
equal in workability two concretes that are subsequently found to
behave differently on the job.

6.1 RESTRICTED VALIDITY OF SINGLE-POINT TESTS

In response to the condemnation of single-point tests, including all the
standard tests, it might be argued that these tests have been found in
practice to be of some use, and that they have given at least some
assistance in the formulation of mix-design procedures and specifications. This argument cannot just be dismissed out of hand but should
be examined further.

There are two, and only two, sets of circumstances under which a
single point test may be useful:
(a) if the effective speed (or shear rate) in the test is the same as that
     of the job in which the concrete is to be used;
(b) if the flow curves of all the concretes to be tested form a pattern
     such that no two of the lines cross each other.

If the first condition applies, the test will at least rank the concretes
in the correct order and the same ranking order will be found on the

job, or, in other words, a concrete classified as more workable by the test will in fact be found to be more workable on the job. This first condition is one that has been tacitly assumed by several workers who have proposed tests that they hope will imitate practical conditions sufficiently closely, and was given explicit expression by Angles[1] in the statement quoted earlier (Chapter 1). A fairly recent example occurs in a Department of Transport model specification for repair concrete[2] whose authors do not refer to any of the British Standard tests but require the use of a Flow-Trough test and a Flow Test for Horizontal Soffit Surfaces. For the latter, they require that the arrangement of reinforcement bars in the test apparatus shall be modified to represent actual reinforcement details.

The defect in these arguments is that it is not at present possible to say what are the effective shear rates in any particular job, and the intended imitation of practical conditions may not be close enough. Even if these objections could be overcome, it would be necessary to have a separate test for each type of job, as seems to have been realized by the writers of the specification just referred to. This means that a strict application of the first condition implies that the only way of assessing the workability of concrete for a particular job is to assess it on the job, so the argument is self-defeating and this first condition is therefore of no practical interest. The relating of workability measurements to job requirements will be discussed in Chapter 13.

The second condition, that of dealing with a series of concretes whose lines do not cross, is of some practical interest. Consider the fan-shaped set of lines shown in Figure 6.3. Since the lines do not cross, the order of ranking obtained from the results of a single-point test will not depend on the (effective) speed at which the test is carried out, so it follows that if a test operates at a speed N_1, say, it will rank

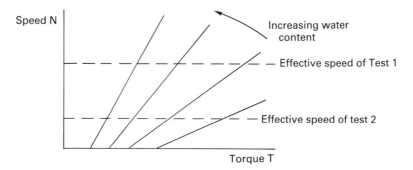

Figure 6.3 Flow curves that do not cross if only water content changes (results of tests with different effective speeds agree in ranking).

correctly the concretes for use in any process operating at any other effective speed such as N_2, say.

Now it happens that such a fan-shaped set of lines is obtained from a series of concretes that differ from each other only in water content and **in nothing else**, so this second condition applies. In the short term this is likely to be the situation on site because the quantity most likely to vary randomly over short periods is the water content, and therefore the standard tests are found to be of some value in that they are in effect acting as a rather crude means of control of water content. However, as soon as anything else changes, trouble may follow. For example, a new delivery of aggregate can throw things awry because it can result in the production of concretes with flow curves that cross the previous set within the important shear rate (or speed) range.

A set of non-crossing lines will also be obtained from a series of concretes that differ only in plasticizer content, and in nothing else, but in this case the lines are parallel to each other, as shown in Figure 6.4. Here too, a single-point test should be capable of classifying the concretes in a correct ranking order and thus of acting as a crude control on plasticizer content. In practice this is not so, partly because the position of a line depends on the time at which the plasticizer was added as well as on the total amount added, but mainly because a batcherman will, on the basis of his subjective assessment of workability, try to correct for any changes arising from variation of the plasticizer addition by adjusting water content. Consequently, two variables require consideration and the situation is no longer simple.

Another way of stating the condition that the flow curves of a set of concretes do not cross is to say that for that set of concretes there is a significant correlation between g and h. In general, this is not so as is shown in Figure 6.5 for a variety of mix compositions and therefore

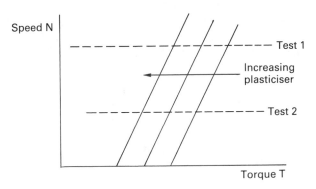

Figure 6.4 Flow curves that do not cross if only plasticizer content changes.

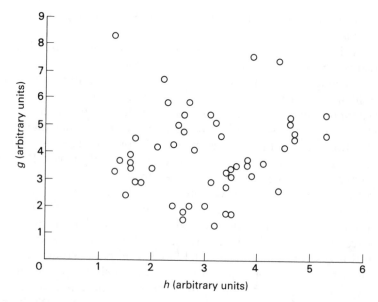

Figure 6.5 Relationship between *g* and *h* (correlation coefficient = 0.19)

it is necessary to measure both *g* and *h*. In the special case of a set of concretes that differ in water content **only**, there is a correlation between *g* and *h* and therefore, in this special case, a single measurement is sufficient to differentiate between the various mixes.

In practice, this is the only case for which a positive correlation exists between *g* and *h* and this fact may be used in the diagnosis of problems on site. In other words, if on site it is found that there is a simple positive correlation between *g* and *h*, it may be deduced that the cause of variation in workability is variation in water content only. If no correlation is found, or a negative correlation, then variation in water content may immediately be ruled out as the main contributor to variation in workability and some other cause must be sought. This will be discussed further in connection with control of concrete quality.

6.2 SIGNIFICANCE OF SINGLE-POINT TESTS

Although no single-point test can provide sufficient information to describe workability, it is obvious that the result from such a test does in some way depend on the workability of the concrete tested, so it is worth considering the nature of this dependence.

The theoretical treatment of the two-point test started from the hypothesis of Metzner and Otto that, although the shear rate in the mixer varies from point to point, one may envisage an average effective shear rate that is simply proportional to the speed of the impeller. It seems reasonable to apply a similar hypothesis to the single-point tests and to suggest that each of them has associated with it an average effective shear rate of an unknown value. As shown earlier, if a measurement is made at only one shear rate, or speed, all that can be assessed is a value of apparent viscosity which will depend on g and h but also on the arbitrarily chosen shear rate at which it is measured. In terms of the two-point test equation therefore, it may be suggested that a single-point test makes some measurement W of apparent viscosity given by

$$W = T/N = g/n + h \qquad (6.3)$$

where n is the value of N characteristic of the single-point test, and is unknown.

In the case of the slump test, the measurement is made on a stationary cone of concrete and it is reasonable to suppose that the rate of shear associated with the test is zero, or, in other words, that the slump test assesses the yield value of the concrete and takes no account at all of the plastic viscosity. If that is so, it would be expected that there would exist a highly significant correlation between slump and g, and such is indeed the case, as has been found by many different investigators. Usually, there is no justification for any relationship more complicated than a simple straight line with an equation of the form

$$S = S_0 - Ag \qquad (6.4)$$

where S is the slump value and S_0 and A are empirical constants, but careful work over a wide range of workabilities may suggest a curved relationship. The earliest results were obtained by Scullion[3] using the original two-point test based on the Hobart mixer and he proposed a relationship of the form

$$S^n = Ag \qquad (6.5)$$

where $n = -0.47$ and $A = 0.007$. This is approximately a square-root relationship. His results are shown in Figure 6.6. More recently, Akashi, Kakuta and Mrimoto[4] reported that the relationship between slump and yield value broke down at slumps in excess of about 150 mm, a fact which they attributed to the difficulty of obtaining meaningful values of slump at such workabilities.

The experimental observations have substantially been confirmed in theoretical studies by Tanigawa, Mori and Watanabe[5], who carried out a computer simulation of the slump test based on the known fact that

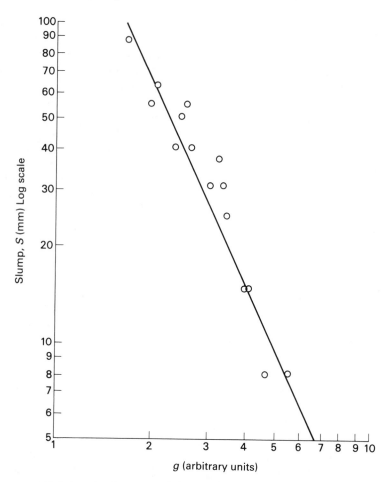

Figure 6.6 Relationship between slump and *g*. (*Scullion*)

concrete conforms to the Bingham model. They showed that the effect on slump of a change in yield value is much greater than that of a comparable change in plastic viscosity, and it is only at low yield value that the latter assumes some importance. They were even able to demonstrate theoretically the asymmetry of the slumped cone that is often observed in practice, and also the retention by the top surface of its original shape.

The fact that over the normal range slump depends only (or at least mainly) on yield value, or *g*, explains immediately why two concretes of the same slump can behave quite differently on a job where the circumstances are such that plastic viscosity, or h, is also important. An example of practical importance is shown in Figure 6.7 which compares the flow curves of two concretes G and L. Concrete G is a concrete with a gravel aggregate and is known to be suitable for piling, while concrete L contains a crushed limestone and is known to be unsuitable because it does not flow satisfactorily. The two concretes have been adjusted to give the same slump, so the slump test says that if concrete G is acceptable then so is concrete L, a result which is known to be untrue. It can be seen from the graph that although the two concretes have the same value of g (yield value), consistent with the fact that they have the same slump, the value of h (plastic viscosity) of concrete L is almost double that of concrete G. Bloomer[6] reports an actual case where some of the batches of concrete for a piling contract were wrongly made with a limestone aggregate instead of with the

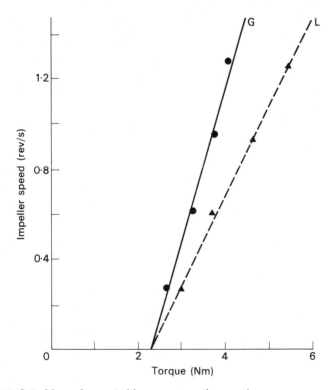

Figure 6.7 Suitable and unsuitable concretes of same slump.

Mix	Slump mm	g Nm	h Nms
G	185	2.3	1.4 suitable for piling
L	190	2.3	2.5 unsuitable for piling

specified gravel; the mistake was detected by two-point workability measurement and confirmed by an examination of concrete spilled round the top of the pile.

The hypothesis that each of the British Standard tests has associated with it an effective shear rate characteristic of the test, expressed in equation 6.3 above, was examined[7] by obtaining from a detailed consideration of Scullion's results an equation that could be tested against the results of several workers who had measured all of slump, compacting factor and Vebe time on the same mixes. The equation obtained was

$$4.55\ V + 255\ S^{-0.47} + 1010\ C = 1000 \tag{6.6}$$

where V is Vebe time in seconds, S is slump in mm, and C is compacting factor. The test consisted of substituting values obtained by other workers in the left-hand side of this equation and comparing the result with the predicted value of 1000. The results were as shown in Table 6.1. The agreement between the actual and predicted values of the function on the left-hand side of equation 6.6 is encouraging and supports the hypothesis on which the derivation of the equation was based. However, in the original paper[7] it was suggested that, for various reasons, the result should be treated with some reservation and this caution was found to be justified by later work by Saeed[8].

Nevertheless, the fact that equation 6.6 shows good agreement with data published by quite independent workers does provide some evidence in support and at least it can be said that the results are consistent with the hypothesis that each of the British Standard tests measures apparent viscosity at an effective shear rate characteristic of the particular test.

The hypothesis does receive further but more indirect support from

Table 6.1 Test of equation 6.6 on other workers' results

Worker	Number of results n	Mean value	Standard deviation	
			of individuals	of mean
Singh[9]	84	1005	32.1	3.5
Dewar[10]	21	986	40.5	8.8
Hughes and Bahramiam[11]	37	903	66.3	10.9
Ritchie[12]	10	984	37.3	11.8
Banfill[13]	146	1007	24.5	2.0
Figg[14]	26	970	45.1	8.9

other sources. Blondiau, Descamps and Stayaert[15] made measurements on the same concretes using the slump, the Vebe, the flow table and the Walz flow test, and reported their results in terms of the correlation coefficients between the tests taken two at a time. If the hypothesis that each test has associated with it an effective shear rate is correct, it would be expected that the results from two tests whose shear rates are the same, or close, would show a higher correlation than would be found for results from two tests whose shear rates are very different. Therefore, it should be possible to arrange the tests used by Blondiau *et al.* in a ranked order, such that the correlation coefficients are greatest for results of adjacent tests, and decrease by increasing amounts as the tests become further apart in the order. Figure 6.8 shows that this is indeed so.

Thus, although the idea that any single-point test effectively measures apparent viscosity at some effective shear rate characteristic of the test, cannot be regarded as proven, it certainly receives quite a lot of direct

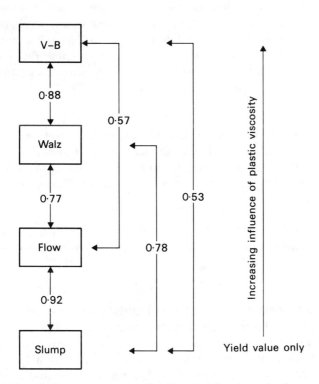

Figure 6.8 Correlations between results of tests taken two at a time. (*Blondiau et al.*)

and indirect support, sufficient for it to be reasonably adopted as a working hypothesis.

This conclusion is of considerable practical importance. In principle, if any two of the three quantities S, C and V are known, it is possible to calculate the third and also g and h and by this means to provide the necessary control of workability. Tanigawa, Mori and Watanabe[5] make a similar point in terms of slump and flow value. In practice, the exercise would be valueless because of the probable errors and other difficulties associated with establishing a suitable system. Of much more practical significance is the fact that the two-point test, whose results are obtained from a simple stirring process, is capable of assessing the behaviour of concrete in completely different processes, and the suggestion follows that it should be possible to label any practical placing process with its own characteristic shear rate. This latter point will be returned to in the discussion of the application of the two-point test in relation to practical job requirements.

There is another very important practical consequence. Several authors have reported that there exist optimum values of some factors, such as fines content, for achievement of maximum workability and have used single-point tests to determine them. Now, if a single-point test assesses apparent viscosity at some effective shear rate equivalent to a speed n characteristic of the test, it is, as shown in equation 6.3 above, measuring a quantity W given by

$$W = g/n + h \tag{6.7}$$

If some change is made in the constitution of a concrete mix there will be an effect on both g and h, and the size and direction of these effects may well differ, as they certainly do for a change in fines content. If the factor changing is denoted by x, the value of W may be written as

$$W = f_1(x)/n + f_2(x) \tag{6.8}$$

where f_1 and f_2 are different forms of function of x. A minimum in W, that is a maximum in workability, will occur when dW/dx is zero, that is, when

$$f_1'(x)/n + f_2'(x) = 0 \tag{6.9}$$

The value of x obtained as a solution to this equation depends not only on the forms of the two functions but also on the value of n. In other words, if a single-point test is used to find, for example, the optimum value of fines content for maximum workability, the actual value of fines content found will depend on the test used to find it; it will be different if a different test is used, and it will be different from the fines content that is best on the actual job.

6.3 REFERENCES

1. Angles J.G. (1974) Measuring workability, *Concrete*, **8**(2), 26.
2. Anon. (1986) Materials for the repair of concrete structures. Department of Transport, Highways and Traffic, Departmental Standard BD 27/86, 15pp.
3. Scullion, T. (1975) The measurement of the workability of fresh concrete, MA Thesis, University of Sheffield.
4. Akashi, T., Kakuta, S. and Mrimoto, T. (1986) Measurement of the physical properties of fresh concrete, *Cement Association of Japan Review*, 178–81.
5. Tanigawa, Y., Mori, H. and Watanabe, K. (1990) Computer simulation of consistency and rheology tests of fresh concrete by viscoplastic finite element method, in H.-J. Wierig (Ed.) *Proceedings of RILEM Colloquium on Properties of Fresh Concrete, held University of Hannover, 3–5 Oct. 1990*, London, Chapman & Hall, 301–8.
   Also: Analytical study of flow of fresh concrete by suspension element method, *ibid*. 309–16.
6. Bloomer, S.J. (1979) Further development of the two-point test for the measurement of the workability of concrete. PhD Thesis, University of Sheffield.
7. Tattersall, G.H. (1976) Relationships between the British Standard tests for workability and the two-point test, *Magazine of Concrete Research*, **28**(96), 143–7.
8. Saeed, Ahmad (1982) Workability measurement with particular reference to the control of concrete production, PhD Thesis, University of Sheffield.
9. Singh, B.G. (1953) Concrete mix design with continuous and gap-graded aggregates, PhD Thesis, University of London.
10. Dewar, J.D. (1964) Relationship between various workability control tests for ready mixed concrete. Technical Report 42.375. London, Cement & Concrete Association.
11. Hughes, B.P. and Bahramiam, B. (1967) Workability of concrete: a comparison of existing tests, *Journal of Materials*, **2**(3), 519–36.
12. Ritchie, A.G.B. (1962) The triaxial testing of fresh concrete, *Magazine of Concrete Research*, **14**(40), 37–42.
13. Banfill, P.F.G. (1977) Discussion of Ref. (7). *Magazine of Concrete Research*, **29**(100), 156–7.
14. Figg, J.W. (1973) Methods of measuring the air and water permeability of concrete, *Magazine of Concrete Research*, **25**(85), 213–9.
15. Blondiau, L.M., Descamps, Ch. and Steyaert, W. (1979) Etude sur les correlations entre les différents tests d'ouvrabilité des bétons frais, Paper No. V in *Rhéologie des Bétons Frais*, Journée de Travail 15 Mai 1979, Université de Liège, Belgian Group of Rheology.

7 Extremely low-workability concretes

The two-point test in its uniaxial form is suitable for medium to high workabilities (about 75 mm slump to flowing concrete) and in its planetary form for low to medium workabilities (about 25 mm to 75 mm slump), and results from the two can be related to each other. Thus the test covers almost the whole range of workabilities on a single scale, an achievement not matched, or even approached, by any other test. It cannot however be used for measurements on extremely low-workability concretes of zero slump or slumps below about 25 mm.

Such concretes, whose water/cement ratios are typically in the range 0.25 to 0.4, are used in the manufacture of products by processes like slip forming, extrusion, or rolling, in which they are subjected to an intense mechanical compacting process, perhaps assisted by the application of vibration or, as in a fairly recent development[1], by the use of a shear-compaction system. The advantages to be gained are that the mould can be removed immediately after compaction, and that the low water/cement ratio results in the development of high strength.

Materials of this type are different in nature from the more normal range of concretes in that they are similar to damp gravelly soils and cannot even be regarded as a continuum before compaction. It is therefore necessary to consider the whole question of workability and its assessment on a different basis. None of the standard tests is suitable, either for study or control, and as yet, no method has been proposed that is fully based on a sound theoretical foundation so that measures of fundamental rheological properties can be obtained.

However, progress in this direction has been made and a promising method has been devised by Paakkinen[2] in the form of his Intensive Compaction Tester, or ICT. His apparatus, which is the subject of patents[3], is shown in the photograph in Figure 7.1 and the principle of the compaction method is shown in Figure 7.2. The essential part

Figure 7.1 The Intensive Compaction Tester. (*Photo courtesey of Invelop OY*)

consists of a cylinder of 100 mm internal diameter which is equipped with opposing pistons that are parallel to each other but inclined at a small angle (about 2°) to the axis of the cylinder. The concrete under test is contained in the cylinder and is compressed under a known applied pressure P, and then, while still under pressure, it is subjected to a shearing motion which is caused by rotating the angle of inclination of the parallel pistons. It is important to note that the pistons themselves do not rotate, but their common angle of inclination to the cylinder axis, α, does, and the effect is to produce movement in the sample as shown schematically in Figure 7.3. The applied pressure can be adjusted between 1 and 7 bar and the speed of rotation between 0.7 and 2.7 rev/s but the values recommended for normal routine testing are 4 bars and 2 rev/s, respectively.

In later developments the apparatus has been modified to a more complex mechanical form but the essential principle remains the same.

During a test, the decrease in volume of the sample is measured continuously by monitoring the mutual approach of the two pistons, and a typical set of curves of density as a function of the number of

Figure 7.2 The Intensive Compaction Tester: schematic (*Paakkinen*).

cycles (i.e. number of rotations) is shown in Figure 7.4. The tilting moment acting on the upper piston may also be measured and a typical example of its value as a function of number of cycles is shown in Figure 7.5.

The quantity of concrete used in the test is such as to give a final height of the compacted cylinder between 100 and 105 mm, and it is claimed that the test takes about 3–5 min. The compacted cylinder may be removed from the apparatus and used for testing either the green material or, after curing, the hardened material for compressive strength or split cylinder tensile strength.

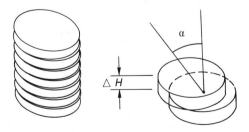

Figure 7.3 Representation of movement in sample (*Paakkinen*).

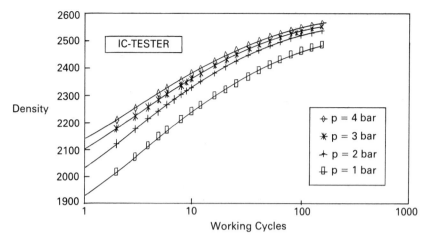

Figure 7.4 Effect of working pressure on density increase (*Paakkinen*).

For routine testing of various mixes comparisons may be made of the densities achieved after 80 cycles with a pressure of 4 bar, as mentioned above, or, for mixes at the higher-workability end, at a pressure of 1 or 2 bar. Before establishing the particular conditions for measurement, Paakkinen[4] had of course carried out experiments to investigate the effects of such factors as the value of the applied pressure, the number of cycles, and the mass of the test sample. He found that the point where the moment/cycle curve begins to drop is

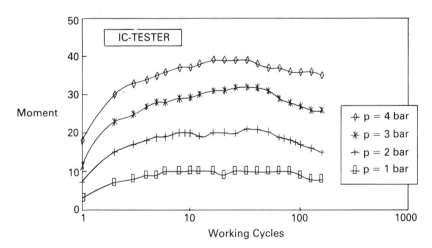

Figure 7.5 Effect of working pressure on working force (*Paakkinen*).

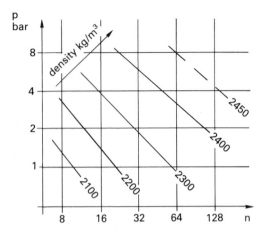

Figure 7.6 Compaction analysis of one concrete sample (*Paakkinen*).

an indicator of the onset of expression of grout from the sample, and he also found that for a given mix, and within limits, pressure and number of cycles were inversely proportional to each other for a given final density. For example, if a particular density is reached after 80 cycles at 4 bar, it will also be reached after 160 cycles at 2 bar. It follows that a plot of pressure against number of cycles, each on a log scale, will for a given density give a straight line with a negative slope equal to unity, and it was found that the lines for different densities were approximately parallel. This finding permits the preparation of a diagram as shown in Figure 7.6.

By taking the measured density of a sample from a production machine, the performance of that machine can be related to a particular combination of the parameters of the IC tester, that is, to a combination of pairs of pressure and number of cycles. This provides a means of comparison of the efficacy of various machines and also establishes for a particular machine the values of the parameters that should be used for experiments on the suitability or otherwise of mixes for use, and for investigations in the design and specification of new mixes. Figure 7.7, prepared as a result of work by Sarja[5], provides this type of information.

While the IC tester has not yet been provided with a sound theoretical basis it does seem to be susceptible to theoretical treatment, and efforts are being made in that direction.

So far, published results of investigations made with this apparatus are scarce, but Juvas[6-8] has used it to study the effects on compactability of no-slump concretes of such factors as water content, age, and the presence of superplasticizers, silt, silica fume and pfa.

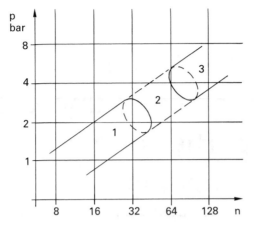

Figure 7.7 Concrete production processes as classified by the ICT. By varying working pressure (P) and work cycles (n), the degree of compaction can be controlled and different production processes simulated (*Saja*).
Region 1 Masonry blocks; concrete pipes; roof tile production
Region 2 Masonry stone; hollow core slabs
Region 3 Paving blocks; roller-compacted concrete

7.1 REFERENCES

1. Schwartz, S. (1984) Practical hollow-core floor slab production below 85 dB(A), *Betonwerk und Fertigteil-Technik*, **(12)**, 807–13.
2. Paakkinen, I. (1986) Intensive compaction tester device for testing the compactability of no-slump concrete. *Nordic Concrete Research Publication No. 5. Dec. 1986*, Norske Betonforening Kronprinsensgate, 0251 Oslo 2, 109–116.
3. United States US 4,794,799; France F 86 04 788; UK Patent Application 8722774.
4. Paakkinen, I. Private communication to G.H. Tattersall.
5. Sarja, A. Technical Research Centre of Finland. Private communication to I. Paakkinen and then to G.H. Tattersall.
6. Juvas, K.J. (1988) The effect of fine aggregate and silica fume on the workability and strength development of no-slump concrete, Master's Thesis, Helsinki University of Technology, 124 pp.
7. Juvas, K.J. (1990) Experiences to measure the workability of no-slump concrete, in *Proceedings of Conference of British Society of Rheology on Rheology of Fresh Cement and Concrete, University of Liverpool, 26–29 March 1990*, London, E. & F.N. Spon, 259–69.
8. Juvas, K.J. (1990) Experiences in measuring rheological properties of concrete having workability from high-slump to no-slump, in *Proceedings of RILEM Colloquium on Properties of Fresh Concrete, University of Hannover, 3–5 Oct. 1990*, London, Chapman & Hall, 179–86.

8 Factors affecting workability: Time and properties of mix components

8.1 INTRODUCTION

The workability of a concrete mix is affected by:
(a) the time elapsed since mixing;
(b) the properties of the aggregate, in particular, particle shape and size distribution, porosity, and surface texture;
(c) the properties of the cement, to an extent that is less important in practice than the properties of the aggregate;
(d) the presence of any cement replacements such as ground granulated blast-furnace slag (ggbs), pulverized fuel ash (pfa) or microsilica (silica fume);
(e) the presence of any admixture such as a plasticizer or retarder or air-entraining agent;
(f) the presence of added fibres;
(g) the relative proportions of the mix constituents.

Thus there are many factors to be considered and the situation is complicated further by the fact that there are interactions between them, that is, they are not independent of each other in their effects. For example, the effect of time is not the same for all mixes, the effect of a change in fines content depends on the richness of the mix, and the effect of a plasticizer may depend on the type of cement in the mix.

Most of these effects have been investigated with the two-point workability test but in some areas the amount of information yet available is still limited, and in others complications have been revealed that will need further work for full elucidation.

The effect on workability of the various factors will be discussed in general terms then in terms of results obtained using the familiar standard tests as well as those obtained from measurements with the two-point test. For the cases in which the two-point test shows that g and h change in the same direction (i.e. both increase or both decrease) when some factor is changed, results from a standard test can be regarded as giving a fair indication of changes that would affect workability as experienced in practice on the job, because here, one is dealing with a set of flow curves that do not cross.

This applies, for example, to the effect of time on an unplasticized mix, or to the effect of a change in water content (only) on any mix. However, if the two-point test shows that g and h may change in opposite directions (i.e. that one of them may increase and the other decrease) the results from a standard test must be treated with considerable reserve, not to say suspicion, because one is now dealing with a set of flow curves that cross each other, and as shown earlier, under those circumstances a single-point test may give indications in the opposite direction to the nature of the change in workability that will be observed on the job. This reservation applies particularly, for example, when attempts are made to find, say, an optimum value of fines content for maximum workability.

8.2 TIME

When cement and water are mixed there is an initial period of very rapid hydration reaction followed by an induction or dormant period (typically of the order of three hours) during which very little reaction takes place. It is the occurrence of this dormant period that makes concrete such a useful material, in that it can be handled, placed and compacted in its fresh state during the dormant period, and then allowed to develop its strength and durability after these processes have been completed. However, workability does decrease with time, and appreciably so in the first few minutes after mixing. It does so because of the occurrence of some further hydration, because of any loss of water by evaporation, and because of absorption of water from the matrix into any particles of aggregate that were not fully saturated at the time of mixing.

Clearly then, the time at which testing is carried out is very important in relation to the time at which the mix was made, and the time at which it is to be used. Older editions of BS 1881 specified a time interval of 6 min between adding the water to the dry mixed materials and completing the workability test, although according to Murdock[1] this time is too short and should be increased to 10 min. The current version of BS 1881 simply specifies that the workability test should be

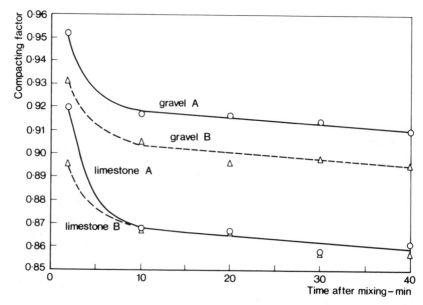

Figure 8.1 Effect of time on compacting factor. (*Murdock*)
——— dry aggregate
– – – soaked aggregate

carried out as soon as possible after sampling and that the time of sampling should be stated. The sample must be protected against evaporation between sampling and testing, and ambient temperature must be recorded. ASTM C172 says the slump test should be carried out within 5 min of completion of sampling.

Murdock's results, shown in Figure 8.1, illustrate the effect of time and, for two different aggregates, the effect of starting with a saturated aggregate or a dry aggregate for the same total water content.

Figure 8.2 shows the effect of time for a typical high-workability concrete, where progressive further additions of water were made in an attempt to maintain workability. Both yield value, g, and plastic viscosity, h, increase with time but the former more rapidly than the latter. Similar behaviour is shown in Figure 8.3. These results were obtained as part of a study of the stiffening of mixes containing superplasticizers but the ones shown here are for very high-workability (flowing) mixes that did not contain a superplasticizer, using four different cements. Once again, yield value increases more rapidly than plastic viscosity. Small but significant differences between the cements are discernible, cement 3 showing the slowest rate of stiffening and

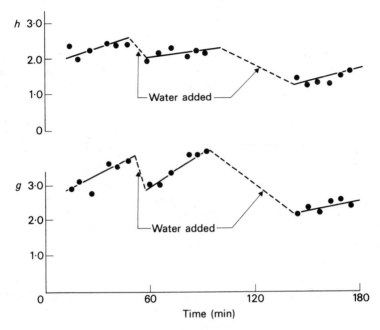

Figure 8.2 Effect of time on *g* and *h* for fresh concrete (MH apparatus). (*Banfill*)

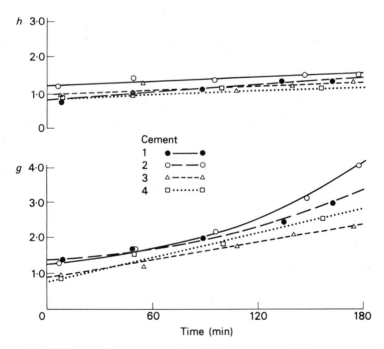

Figure 8.3 Stiffening of concrete made with four cements (MH apparatus; Mix 330 kg/m³ cement, 215 kg/m³ water, 1815 kg/m³ aggregate as Zone 3:10:20 = 42:19:39). (*Banfill*)

cement 2 the most rapid. These two cements had a high C_2S content and a low alkali content respectively, but it is not suggested that any general conclusion about this can be drawn, because the evidence is inadequate.

In a large number of tests on production cements in a standard test mix, Crossley[2] found that the flow curve taken at 20 min was consistently to the right of the one taken at 10 min, that is yield value had increased in the 10 min between tests, and the change was equivalent to a change in water/cement ratio of 0.01. There was no systematic behaviour of plastic viscosity change over this short interval.

Dewar[3] carried out an investigation on the effects of agitation for periods up to five hours upon the properties of ready-mixed concrete. In general, workability as assessed by slump, decreased with time and compressive strength (measured later of course) increased. For a time that varied for different mixes, as shown in Table 8.1, the loss in slump was small and could be regained without loss in subsequent strength simply by adding extra water. Dewar attributes workability loss to evaporation, hydration, absorption, and grinding. The rate of loss is lower for mixes of lower cement content or higher water content, and also for lower air or materials temperature. Some aggregates, particularly limestone fine aggregate or weakly cemented sandstone, are likely to be abraded, so reducing workability, but this effect can be offset by reducing the initial fine/coarse aggregate ratio of the mix.

Loss of workability with time, often referred to as slump loss, may be very important in practice, particularly if the distance between a ready-mixed plant and the site is considerable, or if unforeseen delays occur. The original slump may be recovered simply by adding water, a process that is known as retempering, but the danger is that the added water may cause an increase in water/cement ratio to a value higher than that originally specified and, consequently, a deficit in strength of the hardened concrete. Because of this, the practice is deprecated and, often, specifically forbidden.

Table 8.1 Approximate time of agitation in hours up to which the relationship between slump value and compressive strength was unaltered (Dewar[3])

Aggregate:cement ratio (by weight)	Approximate time of agitation	
	25 mm slump mix	125 mm slump mix
3:1	$\frac{1}{2}$	1
4.5:1	1	2
6:1	2	3
9:1	3	4

However, if the amount of added water does not exceed that lost by evaporation the penalty is avoided, and Dewar's results show that this can be achieved. In fact, if no retempering is carried out the final strength of the hardened concrete will be higher than originally intended, provided (and this is a very important proviso) that the fresh material is still sufficiently workable to be capable of being placed and compacted properly.

Careful investigations reported recently by West[4] lend support to the results of Dewar and to the argument given above. Experiments were carried out on site at temperatures between 8 and 12 °C and in the laboratory at 22 °C, and in each case equations were established to show the effect of known water additions on slump regain and strength loss. From these it was easy to deduce the maximum permissible water addition to improve workability without decreasing the final strength. West himself comments that 'the results presented are limited in scope' but it may be pointed out that the limitation does not extend to the general principle and, as West says, 'indiscriminate retempering has been confirmed as ill-advised (but) . . . limited and considered retempering . . . (may be) a practical solution'. The difficulty in practice is, of course, that of knowing what is the safe limit in particular circumstances, and West recommends further study.

The problem of workability loss is particularly acute in regions of high ambient temperature, and Al-Kubaisy and Palanjian[5] report some work carried out on site at temperatures of 40 to 55 °C. The presence of a plasticizer or superplasticizer also raises additional complications, which are considered later, in Chapter 10.

8.3 PROPERTIES OF AGGREGATES

8.3.1 Particle shape

As a generalization it may be said that the more nearly spherical are the particles of the aggregate, the more workable will be the mix in which they are incorporated, other things being equal. This effect is due to two properties of spheres. First of all there is what may be called the ball-bearing effect, that is simply that it is clearly easier for packed spheres to move relative to each other than it is for particles of angular or awkward shape, even in the dry state. Secondly, for a given mass, the sphere is the shape that has the smallest surface area, so closer approximation of coarse aggregate particles to sphericity means that less mortar is needed for coating them, and also less is needed to fill the voids between them, so more is available to contribute to the general 'flowability' of the mix.

Results conflicting with this argument have been obtained. For example, Erntroy and Shacklock[6] found that, for some of their high-strength mixes of very low water/cement ratio, the compacting factor was greater for concretes containing crushed granite than for similar concretes containing a more nearly spherical irregular gravel.

Nevertheless, it is reasonable to say that, in general, the more nearly spherical are the particles of the aggregate, the better is the workability of the concrete. This may be illustrated by figures taken from *Design of Normal Concrete Mixes*[7]. For a mix of maximum aggregate size 20 mm, the approximate free water content to give a slump in the range 30–60 mm is 180 kg/m^3 if the aggregate is a natural gravel, but the quantity must be increased to about 210 kg/m^3 if the aggregate is a crushed rock, whose particles are much less spherical. For a slump in the range 60–180 mm, the two figures are 195 and 225 kg/m^3.

Lydon[8] illustrates the same point by considering the maximum aggregate: cement ratio that could be used for aggregates of different shapes in a low-workability concrete of 0.5 water/cement ratio and 20 mm maximum aggregate size. His figures are given in Table 8.2. In addition to showing generally the effect of particle shape, Lydon's results indicate that the effect of changing the coarse aggregate only is not so great as that of changing the fine aggregate only. However, this is not true as a generalization and reference to Hughes' mix-design charts[9] shows that for rich mixes the effect of the shape of the fine aggregate may be less than that of the coarse aggregate.

Aggregate shape is usually described only qualitatively by the use of the terms rounded, irregular, angular, flaky, elongated, and flaky and elongated. The application of these terms is illustrated in the photographs in Figure 8.4 and no further explanation of the first three,

Table 8.2 Effect of aggregate particle shape on maximum acceptable aggregate : cement ratio (Lydon[8]) (Low workability mix. water/cement ratio 0.5. Max. size 20 mm)

Shape		Maximum aggregate:cement ratio
Coarse aggregate	*Fine aggregate*	
Rounded	Rounded	7.5 : 1
Irregular	Irregular	5.5 : 1
Angular	Angular	4.7 : 1
Rounded	Irregular	6.5 : 1
Irregular	Irregular	5.5 : 1
Angular	Irregular	5.2 : 1

Figure 8.4 Aggregate particles of various shapes.

which have already been used in Table 8.2, is necessary. A flaky particle is one for which one dimension is considerably less than those in the two perpendicular directions, and an elongated particle is one for which it is considerably greater. For a flaky and elongated particle all three dimensions are considerably different.

It is difficult to describe particle shape in quantitative terms but attempts have been made and, for example, methods are given in BS 812 for the determination of a flakiness index and an elongation index. These two methods involve the use of special sieves with elongated apertures or the comparing of individual particles with special gauges so they can be somewhat laborious, but they are sometimes used in practice. Lees[10] has given a critical assessment of the flakiness and elongation gauges and concludes that they do not satisfactorily fulfil the function for which they were intended. He recommends the use of calipers or adjustable gauges and states that even subjective judgement by the eye of a trained observer is superior to the use of the present gauges.

In BS 812 a flaky particle is defined as one for which the smallest dimension is less than 0.6 times the nominal particle size, and the **flakiness index** of an aggregate is defined as the mass of flaky particles expressed as a percentage of the total mass. BS 882, which deals with natural aggregates for concrete, lays down that the flakiness index of the combined coarse aggregate for concrete shall not exceed 50% for uncrushed gravel and shall not exceed 40% for crushed rock or crushed gravel.

Another quantity that has been used as a measure of particle shape, but is now omitted from BS 812 'because of lack of use', is the *angularity number*. It is determined, for a single-size aggregate, by weighing the quantity of aggregate contained in a standard metal cylinder that has been filled by rodding in a standard manner, and it is defined as

$$\text{Angularity number} = 67 - 100 \frac{M_A}{M_W \rho_A} \tag{7.1}$$

where M_A is the mass of the aggregate, M_W is the mass of water required to fill the container and ρ_A is the specific gravity of the aggregate. Thus $M_W \rho_A$ is the mass of aggregate that would fill the container if there were no voids at all and, since M_A is the mass of aggregate that actually does fill it, the fraction $M_A/M_W \rho_A$ is a measure of the ability of the aggregate to pack, which is known as the **packing fraction**, and is multiplied by 100 to express it as a percentage. It can easily be shown theoretically that the packing factor for closely packed spheres of equal radius is 74% but, to achieve this in practice, each sphere must be carefully placed in the correct position and experiment has shown that the value achieved when equal spheres are placed by methods similar to those used in the test is about 67%. That is why the figure 67 appears in the formula; the angularity number is thus the difference between the packing fractions for equal spheres and the aggregate under test and it represents a measure of the departure from sphericity of the aggregate.

An alternative way of expressing the same thing is to say that the angularity number is given by $V - 33$ where V is the percentage voids in the aggregate as packed.

Kaplan[11] carried out experiments on concretes made with thirteen different coarse aggregates which had angularity numbers from 1 to 10. He obtained a correlation between compacting factor and angularity number, as shown in Figure 8.5 and he also proposed a simple test, which he called the drop test, for measuring angularity. This consisted essentially of carrying out a compacting-factor test on the dry aggregate and, although the percentage voids in the compacted mass so obtained was greater than in the standard angularity test, there was a highly significant correlation between the results of the two. Kaplan also found that changes in flakiness were less important than changes in angularity so far as an effect on compacting factor was concerned. Effects on Vebe time were broadly similar but his results with this method were not as reproducible as those with the compacting factor apparatus so he abandoned it in favour of the latter.

Other measures have been introduced as suggested improvements on angularity number, such as the angularity index of Murdock[1] and the angularity factor of Hughes[12]. Hughes compares the packing of aggregate in a given size range with the packing of spherical glass beads in the same size range. His angularity factor is based on a **loose** bulk density, in contrast to angularity number which refers to a rodded sample, because he considers that, particularly for flaky particles, such

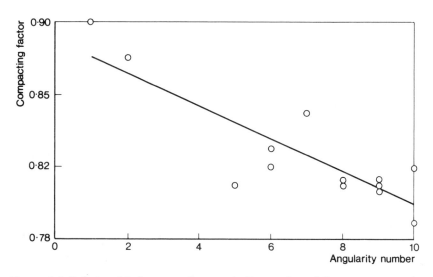

Figure 8.5 Relationship between the angularity number of the aggregate and the compacting factor of the concrete. (*Kaplan*)

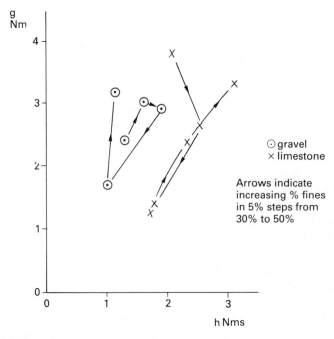

Figure 8.6 Effect of coarse aggregate shape on *g* and *h*. (*after Bloomer*)

packing is more likely to be similar to that in concrete. He has been able to show that the angularity factor of an aggregate can be correlated with the Vebe time of a concrete in which it is used, and it appears as one of the quantities to be considered in his mix-design charts.

There is as yet very little quantitative information on the effect of particle shape on the two parameters, *g* and *h*, of the two-point test, but what there is suggests that, at least for high-workability mixes, the greater change is in *h*. Figure 8.6 shows a plot of some results extracted from Bloomer's work[13], and compares the properties of mixes made with a rounded gravel and a crushed limestone. All mixes contained the same natural sand fine aggregate and all had a cement content of $400\,kg/m^3$; the aggregate:cement ratios at 4.4:1 for the gravel mixes and 4.8:1 for the limestone were slightly different as were the water/cement ratios at 0.42 and 0.47 respectively. Figure 8.6 shows how *g* and *h* change as the percentage fines is altered in 5% steps from 30% to 50%. It can be seen that the two patterns are not dissimilar, that the range of values of *g* is about the same for both, but that the values of h are on the whole higher for the crushed limestone mixes.

A very practical example of this effect and its important consequences has already been given in Chapter 6, where it was shown that of two concretes of the same slump one was suitable for piling and the other was not. The reason is that the unsuitable one had a much higher h value caused by the angularity of the aggregate. Since slump is largely unaffected by h, the slump test cannot pick up this effect.

In *Design of Normal Concrete Mixes* it is pointed out that early mix-design methods used in the UK classified shape as rounded, irregular or angular, but it is now considered that there is in practice insufficient difference between the behaviours of the first two to justify the use of separate classifications. These are the shapes generally associated with uncrushed naturally occurring gravels. The behaviour of angular particles produced by crushing a natural rock is significantly different so, in this method of mix design, classification is simply in terms of uncrushed material and crushed material.

Another important practical point to note is that natural gravels were not all conveniently laid down with a maximum size of 20 mm, which is the arbitrary choice for most of the concrete made in the UK, so particles greater in size are removed by screening and then sent to be crushed before being included in the final product. The proportion of crushed oversize material in a delivery can vary so that a repeat delivery of a nominally rounded gravel may contain a proportion of angular material different from previous deliveries, and there will be an immediate effect on the workability of the concrete produced.

8.3.2 Particle surface texture

The surface texture of aggregate particles may have an effect on the bond between the particle and the matrix, and thus the strength of the hardened concrete, but it has no significant effect on the workability of the fresh concrete.

In the work already referred to, Kaplan[11] measured the surface textures of his thirteen aggregates by the method described by Wright[14] and found that, although there was a wide variation in this property, there was no correlation between it and the compacting factor of the concrete.

On the whole, smooth surfaces are associated with uncrushed aggregates and rough surfaces with crushed ones, although there are exceptions such as smooth crushed flint and rough uncrushed rounded gritstone.

8.3.3 Particle size distribution

It is fairly obvious that the size of particles of an aggregate will have a considerable influence on the workability of the concrete in which it is used. The outdated practice of taking this into account only by considering the relative proportions of coarse and fine aggregates is quite inadequate because different sands will produce quite different results.

An adequate description of particle size can be given only in terms of a distribution showing the proportion of particles within various size ranges. In the case of concrete aggregates, the particles are large enough for the method of sieving to be a satisfactory means of determining this distribution, and the results are presented in the form of the particular type of distribution curve known as a **grading curve**. The method of test is described in BS 812 and a typical result is shown in Figure 8.7. The cumulative percentage passing each of a series of sieves is plotted against the sieve size and the points are joined by straight lines, so that the proportion of material in any given size range is given by the difference in ordinates of the appropriate two adjacent points, or by the slope of the line joining them.

As a first approach, the effect of grading may be considered in terms of what is called the **specific surface**, which is the ratio of the total surface area to the total mass or volume, and is measured in m^2/kg or m^2/m^3. For a given shape of particle, specific surface is inversely proportional to linear dimension so that the finer the particles in a powder, the greater the total area of surface, for a given total mass. In a concrete, this means that the area of surface to be coated and lubricated by finer particles and by cement paste is greater and thus, other things being equal, it would be expected that the finer the fine aggregate, the less workable the concrete. It also suggests that changing the proportion of material in the finer sieve range will have a bigger effect than a similar change at the coarser end. It will be seen later that these conclusions need some qualification.

In any case, it does not follow that the aim is to use as coarse a fine aggregate as possible, with the intended object of achieving as high a workability as possible, because so far no account has been taken of that other most important factor in workability, namely 'stability'. A mix that is deficient in the finer sizes will be 'harsh' and will lack cohesiveness so that segregation will occur and the concrete will no longer be homogeneous. In addition, such a mix will have a very low degree of 'finishability': the production of a satisfactory surface by trowelling or other means will be impossible, or at least difficult. It has been argued that the ideal grading is such that the voids in the

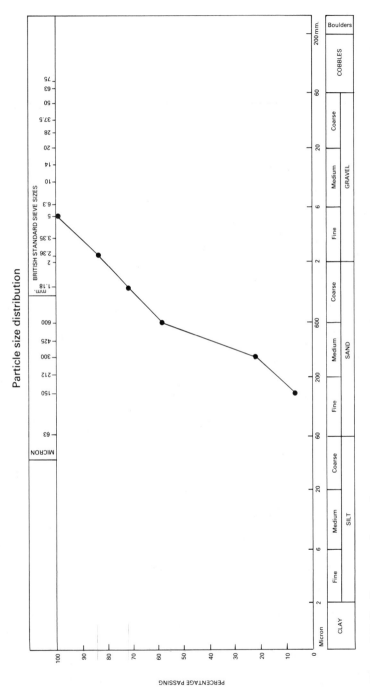

Figure 8.7 Grading curve for sand.

highest size range are filled by particles of the next size range, whose voids are filled by the next and so on.

There have been several attempts to represent the grading of an aggregate by a single numerical value and one of the quantities defined with this intention is the **fineness modulus**. It is calculated by adding up the cumulative percentages by weight of the aggregate retained on the nine sieves from 5 mm to 75 μm and dividing by 100. Since the percentage retained on any sieve is 100 minus the percentage that passes, it can equally well and more easily be calculated by adding up the cumulative percentages passing direct from the grading curve, subtracting from 900 and dividing the result by 100. The definition is rather arbitrary so it is perhaps not surprising that fineness modulus is not a satisfactory measure of particle-size distribution. Murdock and Blackledge[1] attempted to improve on it by calculating what they called the surface index from the grading curve, using empirically determined weighting factors applied to the proportions in the various ranges. Hughes[15] also used weighting factors to calculate his grading modulus but in his case they were obtained theoretically from his definition of the modulus as the surface area per unit volume of spheres which pass the same sieve sizes as the actual aggregate. Whereas Murdock's surface index is not a straightforward measurement of a physical quantity, as he himself emphasizes, but is based on factors empirically determined to fit results of tests in which workability was assessed in terms of compacting factor, Hughes' grading modulus is a meaningful quantity determined from a simple experiment on the aggregate itself which, **after** having been defined, was shown to be usable in predicting the results of the Vebe test.

However, by far the most common way of describing a grading by means of a single figure has been to label it with a number according to which one of several prescribed zones the grading falls entirely or mainly within. In previous editions of BS 882 four grading zones were defined for fine aggregates, such that the grading became progressively finer on passing from Zone 1 to Zone 4, and these categories were used in various mix-design methods. Thus it was usual to refer to an aggregate as, say, a Zone 2 sand, and so on.

It should be realized that to ascribe a sand to a particular zone is to describe it only approximately; two sands in the same zone may exhibit appreciable differences in the behaviour found when they are incorporated in a concrete mix. In addition, the grading curve of a perfectly acceptable sand may wander from one zone to another.

In the current edition of BS 882 the concept of grading zones has effectively been abandoned and wide limits for acceptance for fine aggregates are given, as shown in Table 8.3. This table is accompanied

Table 8.3 Fine aggregate grading limits as given in BS 882:1983

Sieve size	Limits expressed as percentage passing			
	Overall limits	C	Additional Limits M	F
10.00 mm	100			
5.00 mm	89–100			
2.36 mm	60–100	60–100	65–100	80–100
1.18 mm	30–100	30–90	45–100	70–100
600 μm	15–100	15–54	25–80	55–100
300 μm	5–70	5–40	5–48	5–70
150 μm	0–15			

by the statement: 'Good concrete can be made with fine aggregates within the limits specified. In cases where the variability of grading needs to be restricted further for the design of particular mixes or for the adjustment of the fine aggregate content of prescribed mixes, this can be achieved by reference to one or more of the three additional grading limits C, M or F, given in (the) Table'.

The overall limits are so wide that one may question whether it is worth stating them at all, and the additional limits are so wide that a description of a fine aggregate in terms of them alone would not be sufficient for practical specification and mix design. The abandonment of zones in BS 882 has necessitated the introduction of a new way of describing the grading of a fine aggregate in the current version of *Design of Normal Concrete Mixes* and that selected is simply to state the percentage passing the 600 μm sieve. It should be mentioned that BS 882 also gives grading limits for coarse aggregates.

Normally, the gradings of fine aggregate (mainly passing the 5 mm sieve) and coarse aggregate (mainly retained on the 5 mm sieve) are first of all considered separately, but of course it is easy to calculate from them the grading of the combined aggregate as it is to be used in the concrete mix. It is in the adjustment of the relative amounts of coarse and fine aggregate that, in practice, the main control on grading is exercised. This will be considered later in the discussion of the effects of mix proportions.

In addition to the effect of grading, or indeed as part of it, there is an effect of maximum size of the coarse aggregate. Lydon[8] quotes an example as follows. To produce concretes of comparable strengths (e.g. with comparable water/cement ratio of, say, 0.5) and workabilities (e.g. assessed by compacting factor at 0.85), using aggregates of simi-

Table 8.4 Effect of maximum particle size (Lydon[8])

Maximum size of aggregate (mm)	Water content (kg/m³)	Cement content (kg/m³)	Corresponding aggregate : cement ratio
37.5	155	310	6.0 : 1
20.0	169	338	5.5 : 1
10.0	189	378	4.5 : 1

lar particle shape but different maximum size, mixes as shown in Table 8.4 would be needed. As the maximum size of aggregate decreases, it is necessary to add more water to maintain workability and consequently to add more cement to maintain strength. Therefore, it is advantageous to use the largest maximum size possible, but the choice will be restricted by other conditions such as the spacing of reinforcement.

There are indications in work reported by Saeed[16] that for very rich mixes, with an aggregate:cement ratio of 3:1, a change of maximum aggregate size from 10 mm to 20 mm may actually cause a decrease in workability, in that both g and h increased and slump decreased. For mixes of 6:1 aggregate:cement ratio, both g and h decreased and slump increased so, in agreement with Lydon's results, workability increased. In mixes of the intermediate aggregate:cement ratio of $4\frac{1}{2}$:1, there were cases in which g decreased but h increased so, although slump increased, it cannot be said with certainty that workability increased. If the effective shear rate of the job in which the concrete were to be used were to be high enough for h to be the overriding factor, workability would be found to have decreased whereas, if not, workability would be found to have increased.

8.3.4 Particle porosity

The capacity of the aggregate to absorb water may also effect workability. Two otherwise identical mixes, one made with a saturated aggregate and the other with a dry aggregate, will have different workabilities even when the total water content is the same, and also the workability of the latter will decrease with time as water is absorbed from the cement paste. This is illustrated by Murdock's results shown in Figure 8.1.

It should be noted that some methods of mix design are based on total water content and others on free water content.

8.4 CEMENT PROPERTIES

The influence of cement properties on workability is much less important than that of aggregate properties but may have to be taken into account, particularly for rich mixes. It is convenient to consider first the case where one type of cement is substituted for another as when rapid-hardening cement is used instead of ordinary Portland cement, and second, the case where unplanned variations occur between batches of nominally identical cements.

Rapid-hardening Portland cement differs from ordinary Portland cement only in being ground finer to a higher specific surface and in having a higher added sulphate content for control of the hydration of the C_3A. In the ready-mixed concrete industry it seems to be a commonly held belief that, other things being equal, substitution of RHPC for OPC results in reduced workability or, what amounts to the same thing, an increased water demand to maintain a given workability, but there is a lack of hard evidence to support this contention. Experience in the testing laboratories of cement manufacturers with a standard BS 4550 mix containing angular Mountsorrel granite has shown[17–19] that RHPC mixes are **more** workable than their OPC equivalents in spite of the fact that standard consistency tests on **cement pastes** indicate that the water requirement of the former is higher than that of the latter. Evans[20] reports results which show a highly significant positive correlation between slump of the concrete and the specific surface of the cement, as shown in Figure 8.8.

Thus although it seems reasonable to expect that the finer cement with its higher reactivity will be associated with lower workability there is evidence to the contrary. On the other hand, in the course of an investigation on the effects of plasticizers, Al-Shakhshir[21] obtained results on some concretes that did not contain any plasticizer as shown in Table 8.5. The mix composition was cement $300 \, \text{kg/m}^3$,

Table 8.5 Al-Shakhshir's results on unplasticised mixes

| | Specific surface of cement | | | |
| | (360 m²/kg) | | (480 m²/kg) | |
	g	h	g	h
Soaked aggregate	2.34	1.23	2.45	1.21
	2.06	1.59	2.68	1.18
Dry aggregate	1.86	1.38	2.06	1.12
	1.72	1.40	2.04	1.19

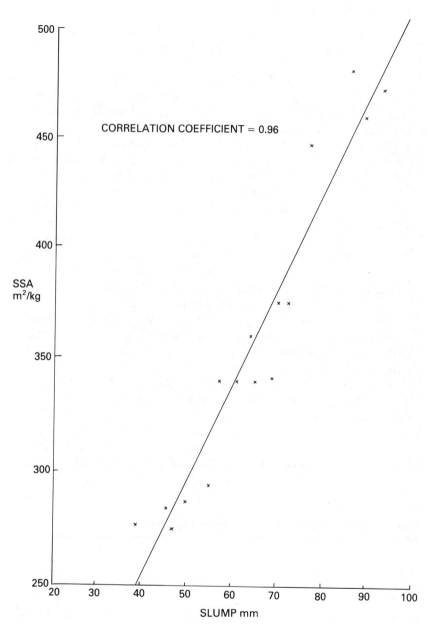

Figure 8.8 Graph of specific surface area v. slump. (*Evans*)

gravel $1132 \, kg/m^3$, sand $755 \, kg/m^3$ (i.e. aggregate:cement 6.3:1, 40% fines). Al-Shakhshir used an ordinary Portland cement and a rapid-hardening cement which were ground from the same clinker to specific surfaces of 360 and $480 \, m^2/kg$ respectively and he used the aggregates in both a dry condition and a saturated condition. The water content was such as to produce a nominal 75 mm slump so it differed slightly for the two cements, being 208 and $211 \, litres/m^3$ respectively, or ratios of 0.696 and 0.704, based on total water content.

Analysis of these results shows that, for h, there is no difference between the mixes made with soaked and dried aggregates, but that the mixes made with RHPC had slightly lower values of h than those made with OPC (significant at about the 0.05 level). However, the slight difference can be explained as the effect of the slightly higher water content of the RHPC mixes. For g, the type of cement has an effect significant between the 0.05 and 0.01 levels, and the state of the aggregate an effect more significant than at the 0.01 level. The value of g is higher for the RHPC mixes, so cannot be explained in terms of the slightly higher water content, and since the two cements came from the same clinker must be attributed to the difference in fineness or the difference in SO_3 content (OPC 2.8%, RHPC 3.4%), or both. However, although these differences have definitely been detected in carefully controlled laboratory experiments, they would be quite insignificant under plant or site conditions, and it is interesting to note that the effect of a change in aggregate (in this case moisture content) is more significant than a change even in the type of cement.

In an investigation carried out by Nkeng[22], OPC and RHPC from nominally the same clinker were obtained from each of four different cement works, with repeat batches taken about a fortnight later, and each batch was used in each of two concretes (nominal C30) made with coarse aggregates of gravel and crushed limestone respectively. Each concrete batch was replicated and the order of experimentation was randomized. Thus the total number of concrete batches investigated was (2 cements) × (4 works) × (2 cement batches) × (2 aggregates) × (repeat) which is equal to 64, and for each of them Nkeng measured slump and the parameters g and h. Each mix contained $300 \, kg/m^3$ cement and the mix proportions were aggregate:cement 6.2:1 with 35.7% fines in the gravel mixes and a slight increase to 37.4% in the limestone mixes. Dried aggregates were used and for each aggregate type preliminary trials were used to find the water content for a slump of about 75 mm, which was then used in all subsequent mixes. (Total water/cement ratio 0.675 gravel, 0.65 limestone).

An analysis of variance carried out on the results showed that for the limestone mixes none of the factors investigated had a significant effect on workability, but for the gravel mixes there were significant interaction terms and it was necessary to carry out further analyses for each of the works separately. These showed that for two of the works none of the other factors investigated had a significant effect while for the other two works there was some indication of an effect of cement type, the RHPC mixes being slightly less workable than the OPC, but the variance ratios concerned were significant only at low levels. The outstanding feature was that for one of the works the effect of cement batch was significant between the 0.01 and 0.001 levels. In other words, in these experiments, the effect of change in cement type was either undetectable or unimportant, and the most important factor causing variation in workability was variation between nominally identical clinkers from one of the works. It should be realised that although this effect has been quite definitely detected in a laboratory experiment the variability in workability arising from this source was quite small and in industrial conditions the effect may well have been swamped by variations arising from other causes.

However, it is generally accepted in the industry that nominally identical cements vary in their 'water demand', which may be defined as the quantity of water required to achieve some arbitrary level of workability in a standard mix. For example, in one investigation it was found that the slump of mixes made to a standard specification (0.65 water/cement ratio) with ten nominally identical cements, from the run of normal production, varied between 65 and 165 mm, with a mean of 110 mm and a standard deviation of 35 mm. Even after allowing for the intrinsic errors of the slump test, these results indicate a significant difference between the cements at the extremes of the range.

Gebauer and Schramli[23] investigated 35 Portland cements from 35 different, worldwide, cement plants and determined water requirements of pastes using the standard Vicat procedure. They concluded that the important parameters of the cement were the C_3A content, the alkali content, and the proportion of particles in the 10 to 30 μm range. Specific surface, except as it was affected by this last factor, was not important.

Of course, the possible variability of cements is recognized by manufacturers. Crossley[2] has reported work with the two-point test in which, by using a standard cement-testing concrete mix but with variable water content, he was able to establish flow curves for intervals of 0.02 water/cement ratio. Then, using his standard strength v. water/cement ratio curves, strength differences were assigned to the flow curves, equivalent to the changes in strength that would be produced if water were added or taken out in a quantity sufficient to bring the flow curve back to that for 0.6 water/cement ratio. Thus if, in subsequent routine testing of production cements all at a water/cement ratio of 0.6, the line for a particular sample was found to be to the right of the line for the works average, a prediction could be made of the strength loss to be expected if more water were added to bring the corresponding concrete back to average workability.

Dimond[24] examined cement/water pastes made from the ten cements mentioned above (i.e. those that gave concrete slumps from 65 to 165 mm). He used both a coaxial-cylinders viscometer and a scaled-down version of the two-point workability apparatus. At a water/cement ratio of 0.35 the yield values ranged from 47 to 72 Pa and the plastic viscosities from 1.3 to 2.2 Pa s. The spread of yield values was thus comparable with the spread of slump values of the concrete, but the correlation between the two was poor, perhaps because of the inadequacy of the slump test. Insufficient material was available for two-point testing of concretes incorporating these cements but it was possible to obtain two further cements which gave yield values and plastic viscosities at the extremes of the range.

In cement pastes of water/cement ratios between 0.36 and 0.53, cement B consistently gave yield values and plastic viscosities double those of cement A, but when they were both tested in concretes of 4:1 and 8:1 aggregate:cement ratio at various water/cement ratios, there was no difference between the g and h values, as shown in Figure 8.9. When experiments on the 4:1 mix were repeated using a different batch of nominally identical aggregate, g increased by about 35%, that is, the unplanned change in aggregate properties had a greater effect than the change from one cement to another of widely different rheological properties. This suggests that the variations in water demand noted by the industry are not always attributable to the cement.

Unfortunately, information on the chemical and physical properties of the cements used by Dimond was not available; but results of a later limited trial involving four cements from four different works suggest that both the compound composition and the fineness could be important. Details are given in Table 8.6. Cements 1 and 2 differ little except in their alkali contents, and the g and h values agree closely. Cement 3, of similar fineness but high C_3A and C_2S, gives higher g and lower h, while cement 4 which is finer and has low C_3A and C_2S content gives lower g and higher h. These limited data suggest that alkali content has no effect but the effects of fineness and proportion of major compounds cannot be separated.

For mixes richer than those referred to so far, the evidence of a connection between cement properties and concrete properties does

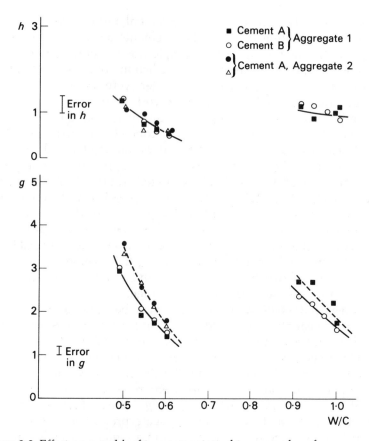

Figure 8.9 Effect on g and h of two cements and two samples of aggregate (*Dimond*)
(MH apparatus; mixes A:C = 4:1, 35% Zone 3 sand and A:C = 8:1, 40% sand.)

become a little stronger. Yeoh[25] carried out a detailed investigation of the relationships between the chemical and physical properties of cement pastes made from 20 different cements, supplied by three different manufacturers, which were considered to exhibit a wide range of water requirements. For 12 of these cements he also carried out two-point workability measurements on mixes of aggregate:cement ratio 3:1, with 40% fines and a water/cement ratio of 0.40. He found significant correlations between the yield value and plastic viscosity of a paste of water/cement ratio 0.365, measured at low shear rates, and the values of g and h, respectively, of the corresponding concretes, as shown in Figures 8.10 and 8.11. However, even for these

Table 8.6 Effect of different cements on g and h of a standard mix (cement 285 kg/m^3, water 185 kg/m^3, aggregate 1930 kg/m^3 as zone 3:10 mm:20 mm = 40:20:40)

	Cement			
	1	2	3	4
C$_3$S	65.9	59.8	53.7	67.8
C$_2$S	7.3	8.5	18.2	5.6
C$_3$A	7.1	8.6	11.6	1.2
C$_4$AF	10.0	11.2	6.7	15.1
SO$_3$	2.3	2.5	2.7	2.3
Alkali*	0.86	0.38	0.55	0.44
Specific Surface (m^2/kg)	344	340	332	385
Slump (mm)	85	55	60	135
g (Nm)	4.17	4.69	5.15	3.45
h (Nms)	1.69	1.60	1.23	1.83

* expressed as % Na$_2$O + 0.658 × % K$_2$O

rich mixes the level of significance of the correlations is not high, and again, the work was carried out under carefully controlled laboratory conditions, using oven-dried aggregates.

Cement is sometimes delivered hot and complaints are made in the ready-mixed concrete industry that there can be serious resulting workability problems, particularly for rich mixes in warm weather. A

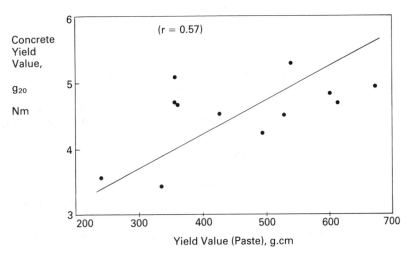

Figure 8.10 Plot of concrete yield value, g_{20}, v. yield value of paste. (*Yeoh*)

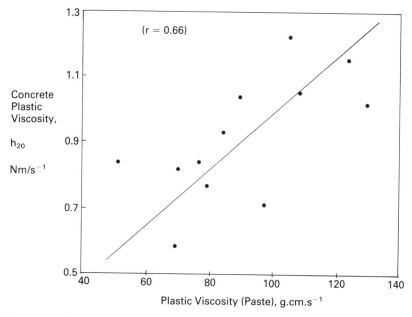

Figure 8.11 Plot of concrete plastic viscosity, h_{20}, vs plastic viscosity of paste. (*Yeoh*)

cement at a temperature of $T°C$ above that of the water and the aggregates can be expected to raise the temperature of a mix with an aggregate:cement ratio of 6:1 and water/cement ratio 0.5, by about $T/10°C$, and of course, the effect for richer mixes and lower water contents will be greater. Keene showed[26] that a variation of temperature over the range from 18 to 71°C had negligible effect as assessed by compacting factor, at periods up to three hours after mixing, but Orr[27], who experimented with cement temperatures of 71, 82 and 93°C, reports a statistically significant effect on compacting factor of the concrete. Orr also found a statistically significant effect of mix temperature which he had investigated at levels of 27, 35, and 43°C.*

Note that in practice cement may be delivered at temperatures in excess of 100°C.

It is possible that the temperature at delivery is important not just in its own right but because it indicates changes in conditions in the process of grinding of the clinker. The grinding of clinker always causes some conversion of gypsum ($CaSO_4 \cdot 2H_2O$) into hemihydrate

---

*See Appendix.

$(CaSO_4 \cdot \frac{1}{2}H_2O)$ and the degree of conversion may be greater than expected. Sharp[28] has reported that in a cement examined by X-ray diffraction he could not detect any peaks due to gypsum whereas those due to hemihydrate were prominent, and similarly, Al-Shakhshir found that in the cements he used the gypsum had almost completely disappeared. The presence of hemihydrate can result in a flash set and perhaps, in less extreme cases, in unusually rapid stiffening. Yeoh, in the work referred to earlier, found a fairly critical minimum gypsum concentration below which the ratio of yield value of a paste to its plastic viscosity increased very rapidly.

More work is needed on the influence of cement properties on workability in both new experimental work and in further analysis of data already available. Present indications seem to be that the important factors in composition are the C_3A content and the quantity and state of the sulphate.

8.5 REFERENCES

1. Murdock, L.J. and Blackledge, G.F. (1968) *Concrete Materials and Practice*, 4th Edn, Edward Arnold.
2. Crossley, A.N. (1983) Reported in Tattersall, G.H., Practical user experience with the two-point workability test, Report BS 74 Department of Building Science, University of Sheffield.
3. Dewar, J.D. (1962) Some effects of prolonged agitation of concrete. London, Cement & Concrete Association, 17pp. Technical Report 42.367.
4. West, R.P. (1990) Concrete retempering without strength loss, in Wierig, H.-J. (ed.), *Proceedings of RILEM Colloquium on Properties of Fresh Concrete, held University of Hannover, 3–5 Oct. 1990*, London, Chapman & Hall, 134–41.
5. Al-Kubaisy, M.A. and Palanjian, A.S.K. (1990) Retempering studies of Concrete in hot weather, *ibid*, 83–91.
6. Erntroy, H.C. and Shacklock, B.W. Design of high strength concrete mixes, in *Proceedings of a Symposium on Mix Design and Quality Control of Concrete, London, May 1954*. London, Cement & Concrete Association, 55–73.
7. Teychenné D.C. *et al.* (1988) *Design of Normal Concrete Mixes*, revised edn, Building Research Establishment, 43pp.
8. Lydon, F.D. (1972) *Concrete Mix Design*, London, Applied Science Publishers, 148pp.
9. Hughes, B.P. (1968) The rational design of high quality concrete mixes, *Concrete*, **2**(5), 212–22.
10. Lees, G. (1964) The measurement of particle elongation and flakiness: a critical discussion of British Standard and other methods, *Magazine of Concrete Research*, **16**(49), 225–30.
11. Kaplan, M.F. (1958) The effects of the properties of coarse aggregates on the workability of concrete, *Magazine of Concrete Research*, **10**(29), 63–73.
12. Hughes, B.P. and Bahramian, B. (1966) A laboratory test for determining the angularity of aggregate, *Magazine of Concrete Research*, **18**(56), 147–52.

13. Bloomer, S.J. (1979) Further development of the two-point test for the measurement of the workability of concrete, PhD Thesis. University of Sheffield.
14. Wright, P.J.F. (1955) A method for measuring the surface texture of aggregate, *Magazine of Concrete Research*, **7**(21), 151–60.
15. Hughes, B.P. (1973) The Vebe test and the effect of aggregate and cement properties on concrete workability, in *Fresh Concrete: Important Properties and their Measurement, Proceedings of a RILEM Seminar, 22–24 March 1973, Leeds*, Leeds, the University, Vol. 2 pp. 4.3-1–4.3-12.
16. Saeed, Ahmad (1982) Workability measurement with particular reference to the control of concrete production, PhD Thesis, University of Sheffield.
17. Evans, D. Private communication, 7 March 1989.
18. Harris, J.T. Private communication, 22 March 1989.
19. Hoult, J. Private communication, 30 March 1989.
20. Evans, D. (1982) The effects of gypsum substitution on the workability of a cement paste and a concrete mix. Project Report. Advanced Concrete Technology Course, Cement & Concrete Association.
21. Al-Shakhshir, A.T. (1988) Workability of plasticised concrete, Dissertation submitted in part fulfilment of requirements for the Degree of MSc (Eng), University of Sheffield.
22. Nkeng, J.E.-S. (1989) The workability of concrete, Dissertation submitted in part fulfilment of the requirements for the Degree of MSc (Eng), University of Sheffield.
23. Gebauer, J. and Schramli, W. (1974) Variations in the water requirement of industrially produced Portland cements, *Ceramic Bulletin*, **53**(2), 161–8.
24. Dimond, C.R. (1980) Unpublished internal report, Department of Building Science, University of Sheffield.
25. An-Keat Yeoh (1982) The effect of cement properties on the workability of concrete, MA Thesis, University of sheffield.
26. Keene, P.W. (1962) An investigation of the effect of cement temperature and ambient temperature on the workability and strength of concrete, DN/17. London, Cement and Concrete Association, 8pp.
27. Orr, D.M.F. (1972) Factorial experiments in concrete research, *Journal of the American Concrete Institute, Proceedings*, **69**(10), 619–24.
28. Sharp, J.R. Private communication, December 1981.

Further Standards referred to

BS 812 *Testing aggregates*

        Pt. 1 1975 Methods for determination of particle size and shape

        Pt. 103:1985 Methods for determination of particle size distribution

        Pt 105:1985 Method for determination of particle shape

           105.1 1985 Flakiness index

BS 882:1983 *Specification for aggregates from natural sources for concrete*

BS 4550 *Methods of testing cement*

        Pt. 3 Physical Tests

           3.5:1978 Determination of standard consistence

9 Effect of mix proportions

9.1 WATER CONTENT

The quantity of water in a mix is usually expressed either as litres/cubic metre or as a water/cement (w/c) ratio by weight, but it is important to note whether the quoted figure refers to total water or to so-called 'free' water, which is the amount after allowance has been made for actual or potential absorption by the aggregate. Only a part of this free water (equivalent to a w/c ratio of about 0.25) is required for reaction with the cement and the rest is present simply to make the mix sufficiently workable for the intended use.

A rough idea of the quantity of water required for various workabilities, as assessed by slump and Vebe time, is given in *Design of Normal Concrete Mixes*[1] and the information is reproduced graphically in Figure 9.1.

For any given mix, an increase in water content always results in an increase in workability in that both g and h decrease monotonously, so that the flow curves for a series of mixes that differ in water content only form a fan-shaped set of lines. A typical set of results, obtained by Al-Shakhshir, has already been shown in Figure 5.9 of Chapter 5, and in Chapter 6 it was shown that a series of this type is the only case for which a single-point test, such as slump, can be expected to give results of any practical value.

For water/cement ratios from about 0.5 upwards the dependence of either g or h on w/c ratio can be satisfactorily represented by a straight-line graph, and it follows that the relationship between g and h themselves is also linear. If the range is extended to lower w/c ratios, down to 0.35 say, the line becomes curved, and Scullion[2] obtained the relationships

$$g = g_0 \exp(-aW)$$
$$h = h_E + h_0 \exp(-bW)$$

These equations mean that both g and h decrease exponentially as water/cement ratio (W) increases but that while g eventually approaches

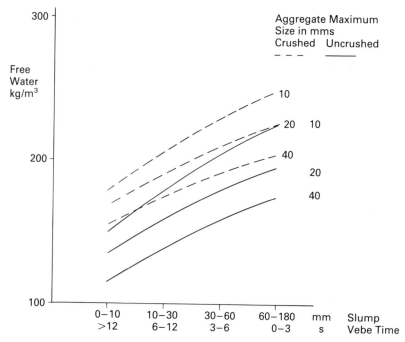

Figure 9.1 Approximate free water contents required to give various slump and vebe times. (From Table 3 of *Design of Normal Concrete Mixes*)

zero, h decays to some non-zero value h_E. In ordinary terms this means that in the extreme the mix becomes a flowing concrete, with a collapse slump, but will still provide some resistance to movement because of the friction and partial interlocking of aggregate; in practice, before this stage is reached quite unacceptable bleeding and segregation would have occurred.

These equations of Scullion, for the greater range of w/c ratio, may seem to be complicated, but the relationship between g and h over a range of w/c ratios of practical interest will still be approximately linear or be represented by a shallow curve.

As stated earlier, this simple relationship between g and h, that is a straight or slightly curved line, is typical of circumstances where only water content varies and it may therefore be used for diagnostic purposes. In other words, if the relationship between g and h of a series of batches is found to be linear, with a positive slope, it can be deduced with confidence that variability in workability is due entirely to variation in water content, and, conversely, if the relationship does not follow this form, some factor other than water content has changed.

This will be discussed again later in connection with the question of quality control of concrete production.

9.2 AGGREGATE : CEMENT RATIO AND FINES CONTENT

The case just considered, of variation of a water content, is a simple one because it can be stated that for all mixes an increase in water causes an increase in workability, but no such generalization can be made for the effects of changing the quantities of aggregate, either in total, or in the relative proportions of coarse and fine. The effects of changes in cement content, aggregate content, and percentage fines, are all interlinked and the effect of altering one of them depends on the values of the others. For example, a change in percentage fines has less effect for rich mixes than it has for leaner mixes.

Moreover, the relationship between workability and any one of the three variables, say percentage fines, is not necessarily monotonous, that is, there may be a minimum in the curve so that an increase in fines may either increase or decrease workability. It is well known in practice that there may be an optimum fines content. If in a highly sanded mix the quantity of fines is reduced, workability may be increased, but a point is reached at which a further decrease in fines causes a decrease in workability. Presumably this is because in the first stage the main effect is to reduce the overall specific surface, and therefore the area to be coated and lubricated, but beyond the optimum this effect is outweighed by the fact that there is insufficient fine material to fill the voids in the coarser material and friction between the larger particles becomes more important.

The effects are even further complicated by the fact that g and h are affected in different ways, and this has the consequence that the optimum fines content for a particular job may be different from the optimum that may be assessed by any of the standard single-point tests.

The very first results, shown in Figure 4.2 of Chapter 4, obtained by the two-point test method using the Hobart mixer illustrate this point. The flow curves for a 1:2:4 mix of various w/c ratios give the typical fan-shaped set, as do those also for a 1:3:3 mix, but those of the latter cross those of the former. This means, for example, for mixes of higher w/c ratios, that one containing 40% fines is more workable than one containing 50% at very low shear rates, but is less workable at somewhat higher shear rates. Without knowing the equivalent shear rate applying on the actual job it is not possible to say which of the two will be the more workable in practice.

It follows that progress can be made only if effects on g and h are studied in detail and then the results used in an attempt to assess the effects on practical jobs. Fairly extensive investigations have been carried out by Scullion, Saeed, and Dimond and Bloomer.

Scullion[2] worked with the early version of the two-point test, based on the Hobart food mixer, and on mixes made from Hoveringham gravel, zone 2 sand, and ordinary Portland cement. He studied all combinations of two maximum sizes (20 mm and 10 mm), three aggregate:cement ratios (3:1, 6:1, 9:1) and three levels of fines content (30, 40, 50%), and for each combination used a range of water/cement ratios. His findings were as follows.

(a)   In general, g increases as fines content increases and the rate of increase is greater for richer mixes or lower water/cement ratios, but at high water/cement ratios (actual value depending on aggregate:cement ratio) increase of fines has little effect.

(b)   There was some evidence of a minimum in some of the g v. fines curves and for the 6:1 aggregate:cement ratio mixes he obtained definite evidence of such a minimum by experimenting at an additional fines content of 33% (see Figure 9.2).

(c)   The effect of fines content on h is even more complicated. For the richest mixes there is little effect at all but for the others the value of h passes through a minimum at a fines content that depends on both w/c ratio and aggregate:cement ratio. Again, in general, the effect of a change in fines content is greater for lower w/c ratios.

(d)   No systematic effect of maximum aggregate size was found.

A replot of Scullion's results is shown in Figure 9.3 for the 6:1, 20 mm mixes and demonstrates clearly that a given combination of values of g and h may be obtained in more than one way. For example, the same pair of values is obtained for a 30% fines 0.6 w/c ratio mix and a 33% fines 0.55 w/c ratio mix, but the effects of changes are very different. In particular, it may be noted that while an increase in fines for the former mix decreases both g and h, either an increase or a decrease in fines for the latter mix increases both g and h.

Saeed[3] used the LM apparatus (planetary motion, H impeller) to investigate a range of mixes similar to those of Scullion. He used an irregular gravel of 20 mm maximum size, a zone 2 sand, and ordinary Portland cement, and made mixes with a:c ratios of 3:1, $4\frac{1}{2}$:1 and 6:1, and 30, 40 and 50% fines. His w/c ratios ranged from 0.35 to 0.70. His results are shown in Figures 9.4 and 9.5. Because he and Scullion worked with different forms of apparatus their numerical results may not be compared directly, but the interrelationships may, and it may be commented that conclusions from the two workers are in general

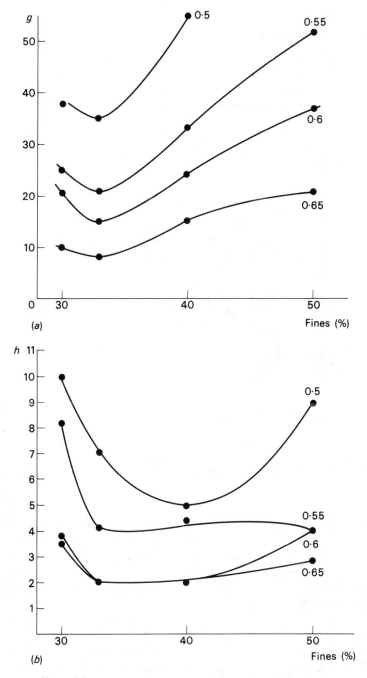

Figure 9.2 Effect of fines content (a) on g and (b) on h. (*Scullion*)

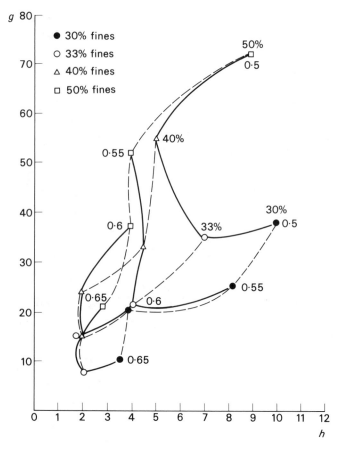

Figure 9.3 Replot of Scullion's results for A:C 6:1, 20 mm mixes. Solid line = constant W/C ratio; broken line = constant fines content.

agreement. Saeed confirmed that for the 6:1 mix there was a marked minimum in the value of g at all w/c ratios. For the other two a:c ratios change in fines content had little effect at higher w/c ratios. He also confirmed that the effect of change in fines on h was much less systematic but there was some evidence of the occurrence of minima at fines contents that depended on w/c ratio.

Bloomer[4] has reported work carried out by himself and Dimond on high-workability mixes of the type used for piling and diaphragm walling, using the MH apparatus (uniaxial rotation, interrupted-helix impeller). They used four different cement contents with a gravel aggregate and one cement content with a crushed limestone aggregate.

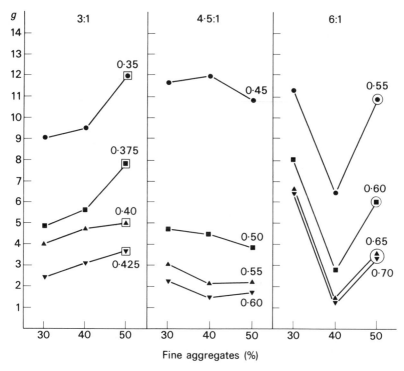

Figure 9.4 Effect on *g* of fines content at different aggregate/cement ratios. (*Saeed*)
□ = average of 4 results
○ = average of 3 results
All other points average of two results.

At each cement content they covered a range of fines contents which had been chosen to be in the neighbourhood of the minimum of the graph of percentage voids plotted as a function of fines content, determined by the method due to Kempster[5]. Their results are shown in Table 9.1 and Figure 9.6. Note that the free w/c ratio increases as cement content increases to keep the mixes within the same practical workability range, and that each mix contained 0.28% by weight of cement of Flocrete N, a conventional lignosulphonate plasticizer.

As found by many other investigators, there is a highly significant correlation between slump and *g*, so, in conformity with that result, for each cement content the slump v. fines curve is a mirror image of the *g* v. fines curve. The remarkable thing is that both these curves show three turning points, i.e. two minima. This effect has not been reported before, although other workers besides Scullion and Saeed

Table 9.1 Values of g and h for various high-workability mixes (*Bloomer*)

Mix number	Nominal cement content (kg/m³)	A/C	Fines (%)	Free W/C	Time* (min)	Slump (mm)	Two-point results			
							Time* (min)	Corr coeff.	g (Nm)	h (Nm s)
Gravel mixes										
F1	520	3.1	30	0.39	11	200 c	16	0.984	2.05	1.14
F2			35		10	175	16	0.995	2.47	1.07
F3			40		10	185	16	0.988	2.54	1.07
F4	460	3.8	25	0.38	13	88	16	0.998	4.73	1.48
F5			30		10	164	16	0.998	3.53	1.68
F6			35		10	134	12	0.990	4.17	1.54
F7			40		14	190	17	0.999	2.05	1.21
F8			45		15	sub c	16	0.994	1.62	1.48
F9			50		13	158	16	0.994	2.97	1.41
F10	400	4.35	30	0.42	13	185 c	16	0.999	2.40	1.34
F11			35		13	159	17	0.995	3.04	1.61
F12			40		10	167	14	0.998	2.89	1.88
F13			45		13	sub c	16	0.994	1.69	1.01
F14			50		10	169	16	0.996	3.18	1.14

F15	350	5.11	30	0.47	13	137 S	16	0.997	3.81	1.74
F16			35		13	115	16	0.985	3.60	2.15
F17			40		13	118	16	0.998	4.17	1.81
F18			45		13	191	16	0.940	2.33	0.87
F19			50		13	63	16	0.994	6.78	0.87
Limestone mixes										
F20	400	4.68	30	0.47	10	197 c	18	0.980	3.81	2.08
F21			35		13	168	17	0.997	2.61	2.55
F22			40		13	sub c	17	0.997	1.41	1.81
F23			45		9	sub c	18	1.000	1.27	1.74
F24			50		10	170	16	0.999	2.40	2.35
F25			55		10	180	19	0.998	3.32	3.15

* After first addition of mixing water.
S = shear slump.
c = collapse slump; sub c = subjectively judged to be a collapse slump.

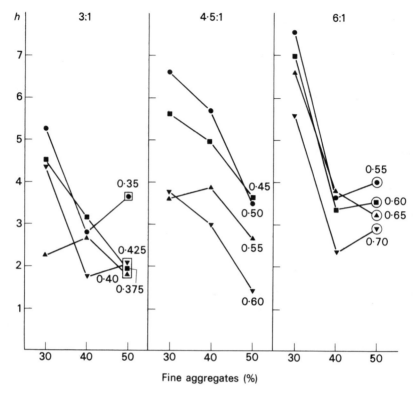

Figure 9.5 Effect on *h* of fines content at different aggregate/cement ratios. (*Saeed*)
□ = average of 4 results
○ = average of 3 results
All other points average of 2 results.

have reported single minima, and in fact if anyone had discovered the double effect on the basis of the slump test he would in all probability have ignored it because differences are within the generally accepted error of that test. The effect is undoubtedly real because it has been detected by the two-point test and also because the curves obtained at different cement contents are similar in shape. The effect of fines content on *h* is less than on *g* and the range of *h* decreases as cement content increases. Only one cement content (400 kg/m³) was used with crushed limestone and a w/c ratio higher than that for the comparable gravel mixes was required to give workability in the same range of high slump values. The *g* v. fines graph had only one turning point (a minimum) while the *h* v. fines graph was similar to those of the gravel mixes but the range of *h* values was greater.

Apart from a few statements of the type that the effect of a change

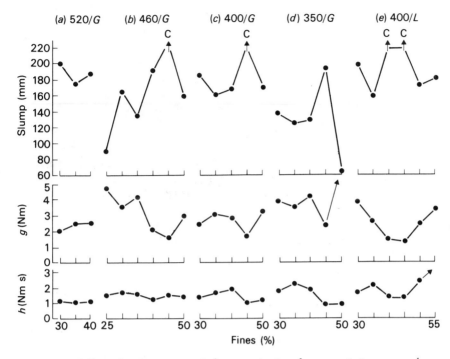

Figure 9.6 Effect of cement content, fines content and aggregate type on g, h and slump values. (*Bloomer*)

in fines content is less at high cement contents or low water/cement ratios, it is not possible to generalize satisfactorily about the effects on workability of changes in mix composition because the behaviour is so complicated. There are interactions between variables, that is, the effect of altering one factor depends on the values of the other factors so any rule becomes specific to one particular starting mix.

The possible ways of dealing with this complex situation will be discussed later in the consideration of control of concrete quality.

9.3 SEGREGATION AND BLEEDING

It has already been stated that, in a full consideration of workability, account must be taken of the stability of the mix, that is, its ability to resist segregation and bleeding. Figure 9.7, which is due to Lynsdale[8], shows the top and bottom surfaces, after grinding, of two discs that were cast on a vibrating table from two batches that differed only in water content. The mix had an aggregate cement ratio of 5.8:1 with 40% fines, but the water/cement ratio of one batch was 0.5 and of

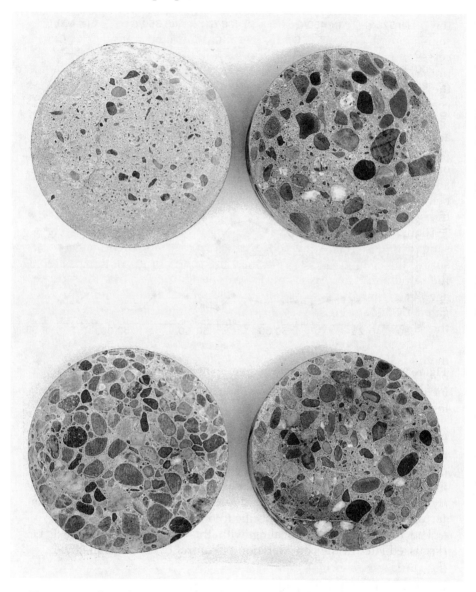

Figure 9.7 Effect of increased water content on segregation. (*Photo courtesy of Dr C. Lynsdale*)

0.65 W/C   0.50 W/C

Top   Bottom

the other 0.65. The marked effect of the increase in water content on segregation is obvious. Clearly, it is important that during all the processes of mixing, transporting and placing the concrete must retain its uniformity of composition.* Some mention has already been made of the factors that are important in this connection but it is helpful to consider them together.

Popovics[6] has published a comprehensive review on segregation and bleeding in which he distinguishes between two types of segregation, internal and external. The former is settling of the coarsest or heaviest particles towards the bottom, whereas the latter is a separation of the coarser particles from the main body and is usually caused before consolidation by external forces such as improper methods of transport.

Tendency to internal segregation may be tested by subjecting a sample to a standard method of consolidation and then either measuring the change in the position of the centre of gravity, or splitting the sample into an upper and lower half and determining the coarse-aggregate content of each. Density distribution has also been investigated by measuring the attenuation of gamma rays passed through the sample.

Tendency to external segregation has usually been estimated by methods that involve dropping the concrete under standard conditions, but Walz[7] proposed a test in which the concrete was passed over an inclined riddle and the increase in coarse-aggregate content, due to loss of mortar through the apertures of the riddle, was measured.

Segregation is usually harmful and it is much easier to prevent it than to repair the damage it causes. Popovics lists the following as factors that contribute to an increased tendency to segregation:

(a) increase in maximum particle size over 25 mm;
(b) increase in the quantity of largest aggregate fractions;
(c) increase in the specific gravity of coarse aggregate compared with the specific gravity of fine aggregate;
(d) decrease in proportion of fines;
(e) decrease in cement content;
(f) unfavourable change in particle shape;
(g) change in water content to make the mix either too dry or too wet.
Addition of an air-entraining agent or a finely divided pozzolanic material reduces the tendency to segregation.

Bleeding is the appearance of water at the surface of concrete after consolidation and may occur either as 'normal bleeding' or as 'channelled bleeding'. The former is a uniform seepage over the whole

---

*See Appendix.

surface, which in small quantities is not harmful, and may even be an advantage if the procedure for curing is inadequate, but the latter occurs in localized channels and is always harmful. Numerical characterization of bleeding is in terms of three quantities: the bleeding capacity, which may be defined as the volume of water exuded per unit volume of mix; the rate at which bleeding occurs; and the duration of bleeding or the time at which bleeding becomes negligible. These quantities can be measured by decanting the supernatant water or by collecting it over carbon tetrachloride, with which it is immiscible. Browne and Bamforth's pressure-bleed test, used in relation to pumping, is described later.

Generally the bleeding rate can be reduced by:
(a)　increase in fineness of the cement;
(b)　increase in cement content;
(c)　addition of pozzolanas;
(d)　decrease in water content;
(e)　addition of an air-entraining agent.
These considerations apply also to total bleeding capacity, with the notable exception that an increase in cement content usually results in an increase in total bleeding capacity because of the reduced bridging effect of the aggregate particles.

In experiments on cement pastes, Suhr and Schöner[9] demonstrated the importance of the concentration and the form of the sulphate present. Their graph of total bleed water as a function of hemihydrate concentration showed a peak whose position occurred at higher concentrations for higher temperatures. In other words, at the temperatures they used (5, 20 and 30 °C) there was a particular hemihydrate comcentration which gave maximum bleeding and at higher or lower concentrations the tendency to bleed was less. This decreasing tendency to bleed, on each side of the peak, was explained as being due to the build-up of structure caused by the formation of additional reaction products; on one side, a lack of easily soluble sulphate encouraged the development of monosulphate, while on the other side, excess sulphate tended to recrystallize.

It might reasonably be expected that, for a given mix specification, there should be a simple relationship between the bleeding of the concrete and the bleeding of the cement-paste fraction, so that the former could be predicted from the latter. Such a correlation was found by Bielak[10] but, although it was significant at about the 0.001 level, there was a wide spread of the points about the line and Bielak states that only 30% of the bleeding of the concrete could be explained in terms of water segregation of the cement.

9.4 SEGREGATION DURING TESTING

The possibility of the occurrence of segregation during the carrying out of a test should be borne in mind. In the case of the two-point test it is sometimes noticed that there is a relative concentration of larger particles on the surface of the sample towards the end of a test, and a concentration of finer particles is found in the material below the impeller when the bowl is emptied. For the great majority of mixes there is no problem and even when some segregation is observed the composition of the material in the central zone is still representative. Wallevik and Banfill[11] examined samples taken from various positions in the bowl after various times of mixing and concluded that, during a single test taking a few minutes, 'the aggregate grading in the zone sheared by the impeller is unchanged'. Similar later, unreported, experiments have confirmed this finding, but prolonged and unnecessary shearing of the sample should be avoided.

The impeller rotates in a sense such as to lift the sample of concrete and that is intended to oppose any tendency to segregate. If serious segregation does occur it is likely that the mix is in any case an unsatisfactory one that would be likely to segregate on the job. If segregation sufficient to affect the results does occur during the test, it will of course be betrayed by a change in the torque (or pressure) reading obtained at a particular speed. Ellis and Wimpenny[12] have exploited this phenomenon to suggest a means of using the two-point test to investigate the segregation tendency of various mixes.

Their experiments covered 15 mix designs including cement contents of 200, 300, and $400\,kg/m^3$, replacement levels of 0, 40 and 70% for each of two different slags, and replication of each mix. For each of the 30 batches the flow curve was obtained using the apparatus in the LM form (planetary motion, H impeller) and then immediately a repeat measurement was made of the torque at the top speed used in determining the down-curve. The difference between the first and second readings of the torque at the top speed was designated T_c.

Stability of each batch was then assessed in two ways.

(a)   The fresh concrete was examined subjectively for its tendency to bleed and each batch was assigned a number, the bleeding mark, B_m, on a scale ranging from 1 representing a low tendency to 10 representing a high tendency.

(b)   The hardened concrete was examined for homogeneity by measuring the transit times of an ultrasonic pulse at different depths in a $500 \times 100 \times 100\,mm$ beam. The difference in the transit times at depths of 25 mm and 75 mm, measured in the direction of

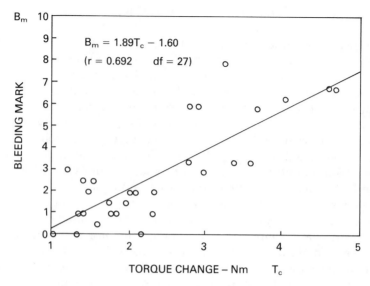

Figure 9.8 Bleeding mark v. torque change. (*Ellis and Wimpenny*)

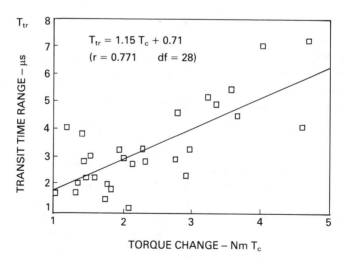

Figure 9.9 Ultrasonic pulse transit time range v. torque change. (*Ellis and Wimpenny*)

casting, was taken as a measure of inhomogeneity and designated T_{tr}.

Figures 9.8 and 9.9 show the results. There is a highly significant correlation, at better then the 0.001 level in each case, between torque change T_c and bleeding mark B_m, and between T_c and transit time difference T_{tr}, so Ellis and Wimpenny conclude that the two-point test is a useful method for assessing stability of a mix. Additional support for this conclusion was obtained from further experiments in which flow curves were determined at various times after mixing and the results correlated with ultrasonic pulse velocities measured at eight different depths in $150 \times 150 \times 450$ mm columns.

Wallevik and Gjørv[13] have also assessed segregation by carrying out a repeat torque measurement at the top speed but have not related the results to any measurement other than in the two-point test itself. For a mix of 140 mm slump they found appreciable segregation had taken place in the test bowl even after 7 min and they consider that the time of testing should not be greater than about 2 min. This can be achieved even with the simple pressure-gauge method of measurement of torque but Wallevik and Gjørv have made it much easier by using a pressure transducer and an automatic tachometer both coupled to a recorder.

9.5 REFERENCES

1. Teychenné, D.C. *et al.* (1958) *Design of Normal Concrete Mixes*, revised Edn, Building Research Establishment, 43pp.
2. Scullion, T. (1975) The measurement of the workability of fresh concrete, MA Thesis, University of Sheffield.
3. Saeed, Ahmad (1982) Workability measurement with particular reference to the control of concrete production, PhD Thesis, University of Sheffield.
4. Bloomer, S.J. (1979) Further development of the two-point test for the measurement of the workability of concrete, PhD Thesis, University of Sheffield.
5. Kempster, E. (1969) Measuring void content: new apparatus for aggregates sands and fillers, *Contract Journal*, **228**(4683), 409–10. Also Building Research Station Current Paper 19/69.
6. Popovics, S. (1973) Segregation and bleeding in *Fresh Concrete: Important Properties and their Measurement*, *Proceedings of a RILEM Seminar 22–24 March 1973, Leeds*, Leeds, The University, Vol. 3 pp. 6.1–1 to 6.1–36.
7. Walz, K. (1939) Die Verarbeitbarkeit des Betons, *Beton und Eisen*, **38**(21), 327–31, and **38**(22), 337–9.
8. Lynsdale C.J. Private communication, March 1990.
9. Suhr, S. and Schöner, W. (1990) Bleeding of cement pastes, in *Proceedings of RILEM Colloquium on Properties of Fresh Concrete, University of Hannover 3–5 Oct. 1990*, London, Chapman & Hall, 33–40.
10. Bielak, E. Testing of cement, cement paste and concrete, including bleeding. Part 1: Laboratory test methods, *ibid.* 154–66.

138 Effect of mix proportions

11. Wallevik, O. and Banfill, P.F.G. (1983) Internal Report, University of Liverpool.
12. Ellis, C. and Wimpenny, D.E. (1990) The assessment of mix stability using the two-point test, in P.F.G. Banfill (Ed.) *Rheology of Fresh Cement and Concrete, Proceedings of International Conference British Society of Rheology, University of Liverpool, March 1990.* Ed. F.N. Spon, 1990, 281–291.
13. Wallevik, O.H. and Gjørv, O.E. (1990) Modification of the two-point workability apparatus, Paper to be published. Draft copy communicated privately to G.H. Tattersall, March 1990.

10 The effect of chemical admixtures

Admixtures may be used in concrete mixes for a variety of purposes including accelerating or retarding the set, waterproofing, or entraining air, and some of these will have an effect on workability. However, some admixtures, known as plasticizers or superplasticizers, have been developed specifically to improve workability for placing and compacting, and others as pumping aids to avoid blockages and to widen the range of pumpable mixes. Table 10.1 summarizes the most common types of admixture and gives an indication of the effect of each on the rheological properties of the fresh concrete. Several books and reviews have dealt with the types, mode of action, and technology of these materials[1-4].

10.1 PLASTICIZERS AND SUPERPLASTICIZERS

The principal types of plasticizer and superplasticizer are the lignosulphonates, which are obtained from the waste liquor from wood pulping, and the synthetic resin types which are aqueous solutions of naphthalene formaldehyde sulphonates, or melamine formaldehyde sulphonates. All of them act by being adsorbed at the surface of cement particles in water to cause deflocculation and of course, all of them produce an increase in observed workability.

They are used in three ways in practice:
(a) by adding to a mix whose composition is otherwise unaltered, so giving increased workability with the same long-term strength and durability;
(b) by adding to a mix with reduced water content, to get the same workability as the original mix but with higher strength;
(c) by adding to a mix in which both water content and cement content are reduced, to get the same workability and strength as

Table 10.1 Concrete admixtures – the most common types

Admixtures	Effects	Typical materials	Advantages/uses	Disadvantages	Effect on workability
Accelerators	(a) More rapid setting	Sodium aluminate Sodium silicate Lime Potassium hydroxide	Facilitate sprayed concreting		Increased rate of change with time
	(b) More rapid strength development	Calcium chloride Calcium formate Sodium nitrite	Normal rate of strength development at low temperature Shorter times to striking formwork	Possible cracking due to heat evolution Risk of corrosion of embedded metal	
Retarders	Delay setting	Hydroxycarboxylic acids Sugars	Maintain workability at high temperatures Reduce rate of heat evolution Extend placing times	May promote bleeding	Initially may be significant Reduced rate of change with time
Water-reducers (Plasticizers)	Increase workability	Calcium and sodium lignosulphonate	(a) Higher workability with strength unchanged (b) Higher strength with workability unchanged (c) Less cement for same strength and workability	Retardation at high dosages Risk of segregation Premature stiffening under certain conditions	Very significant effect in all three uses
Superplasticizers (super-water-reducers)	Greatly increase workability	Sulphonated melamine-formaldehyde resin Sulphonated naphthalene-formaldehyde resin Mixtures of saccharates and acid amides	As for water-reducers, but over wider range Facilitate production of flowing concrete	Risk of segregation Time of addition to mix may be important May increase rate of loss of workability	Very significant effect in all uses May modify rate of change with time

Accelerating water-reducers	Increase workability with faster strength gain	Mixtures of calcium chloride and lignosulphonate	As for water-reducers, with faster strength development	Risk of corrosion of embedded metal	Very significant initial effect. Increased rate of change of time
Retarding water-reducers	Increase workability and delay setting	Mixtures of sugars or hydroxycarboxylic acids and lignosulphonate	As for water-reducers, with slower loss of workability		Very significant initial effect. Decreased rate of change with time
Air-entraining agents	Entrain small air bubbles into concrete	Wood resins, Fats, Lignosulphonates	Increase durability to frost without increasing cement content and heat evolution	Careful control of air content and mixing time necessary. Workability increase may be unacceptable	Significant increase
Waterproofers	(a) Prevent water from entering capillaries of concrete	Soaps, Butylstearate, Petroleum waxes	Reduce surface absorption, staining		Probably negligible
	(b) Reduce permeability of concrete to water	Very fine particulate minerals	Facilitate production of watertight structures without using very low water/cement ratios		Slight
Pumping aids	Prevent segregation of mix during pumping to avoid blockage	Polyethylene oxide, Cellulose ether, Alginates, Alkylsulphonate	Widen the range of mixes suitable for pumping		Significant increase

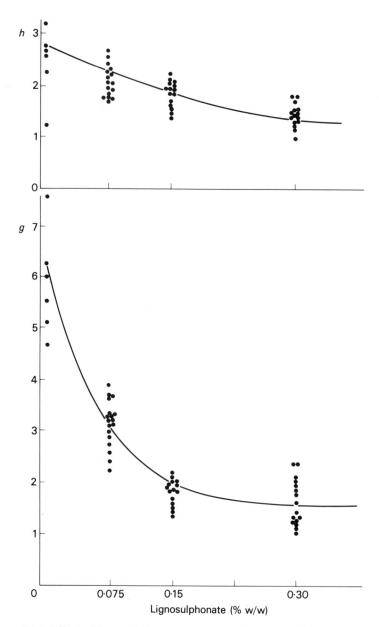

Figure 10.1 Effect of lignosulphonate concentration on g and h.
(MH apparatus; mix 350 kg/m^3 cement, 227 kg/m^3 water, 1850 kg/m^3 aggregate
as Zone 3:10:20 mm = 40:18:42) (*Waddicor*)

the original mix but more economically because of the saving in cement. Of course, this can be done only if durability requirements are still met.

Because of the second and third uses, these substances are sometimes referred to as water-reducing agents.

The presence of retarding impurities in most lignosulphonates limits the concentration that can be used, while the synthetic materials, without retarders, can be used to higher concentrations and thus achieve greater gains in workability or reduction in water content. At very high workabilities segregation may occur unless the mix composition is modified to ensure that there is a sufficient quantity of fine particles present[5].

Early work by Scullion[6] showed that a lignosulphonate has a greater effect on g than on h and this was confirmed later by Waddicor[7] on a series of commercial lignosulphonates. His results for eighteen different admixtures are shown in Figure 10.1. He also found that increased molecular weight of lignosulphonate gave a greater reduction in g but not in h, that degree of sulphonation had no effect on either, and that at a given molecular weight sodium lignosulphonates had a greater effect on g than had calcium lignosulphonates.

The so-called superplasticizers, which are used at higher concentrations, must be free of retarding impurities, so they must be either the synthetic resins referred to above or they must be ultra-pure lignosulphonates. They are used to increase workability to such an extent as to produce 'flowing' concrete, whose yield value has been reduced so far that little or no vibration or other compacting effort is needed, and it has been suggested[8] that for these self-levelling concretes g should not be greater than about 2 Nm. Banfill[9] thinks that the best mixes are probably in the range $g = 1 - 1.5$, $h = 0.4 - 1.0$.

Rixom and Waddicor[10] found that the ultra-pure lignosulphonates compare favourably in performance (and price) with the synthetic resins, as shown in Figure 10.2. Again, it can be seen that the major effect of the admixture is on g, with little or no effect on h. Banfill[9] studied a naphthalene resin and a melamine resin with four different cements and he also found a major effect on g, but some indication of a slight increase in h followed by a decrease as concentration increased. This is shown in Figure 10.3. Some confirmation of this latter result is afforded by Bloomer[11] who studied the effects of a standard dose of a commercially available melamine resin admixture on mixes of three cement contents. As shown in Table 10.2, g always decreased but at each cement content there was a slight increase in h above that of the control mix. Bloomer also showed that the effect on h can be reversed by altering the sand content, as shown in Figure 10.4. There

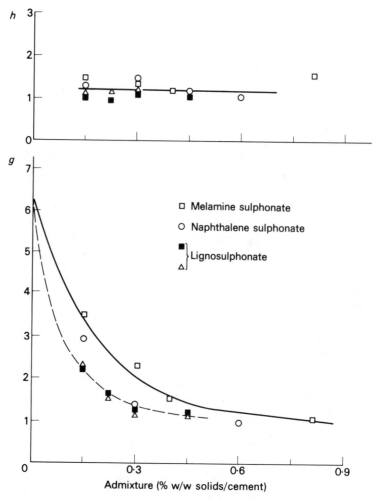

Figure 10.2 Effect of superplasticizers on g and h. (MH apparatus, mix as Figure 10.1) (*Rixom and Waddicor*)

is little effect of sand content on the change in g but moving from 35% to 45% sand alters the change in h from positive to negative.

Although there are some exceptions, it may be said as a general practical guide that the major effect of adding a plasticizer or super-plasticizer is on g, with little or no effect on h. This means that a series of mixes of increasing plasticizer content but with other factors constant, will give a series of parallel flow lines as shown in Figure 6.4 of Chapter 6, because h, the reciprocal slope, remains constant while g,

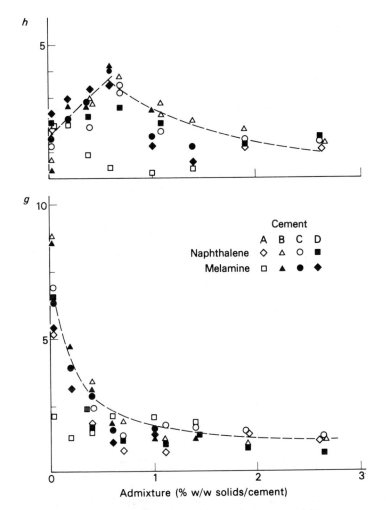

Figure 10.3 Effect of superplasticizer concentration on g and h.
(MH apparatus; mix 279 kg/m³ cement, 195 kg/m³ water, 1890 kg/m³ aggregate as Zone 3:10:20 mm = 42:19:39) (*Banfill*)

the intercept, decreases. The appearance of such a pattern in practice is an indication that the main factor causing variability of the concrete is one associated with the plasticizer. This will be discussed in more detail later in connection with control of concrete production.

Another consequence of the fact that plasticizers affect only g, whereas water affects both g and h, is that it is not possible to alter a mix to obtain the same workability at a lower water content by adding a plas-

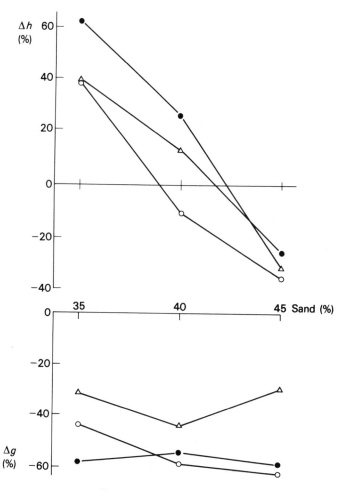

Figure 10.4 Effect of three superplasticizers on g and h for mixes of different sand content. (MH apparatus; mix 365 kg/m^3 cement, 182 kg/m^3 water, 1825 kg/m^3 aggregate to 20 mm) (*Bloomer*)

Control mixes:	sand (%)	g	h
	35	4.4	1.4
	40	5.1	1.3
	45	3.8	2.1

ticizer, or, in other words, it is not possible to produce two mixes of the same workability that differ only in plasticizer content and water content. This is contrary to what is generally believed by those who use a plasticizer for the second of the uses listed at the beginning of this chapter. Figure 10.5 illustrates the point by showing the results

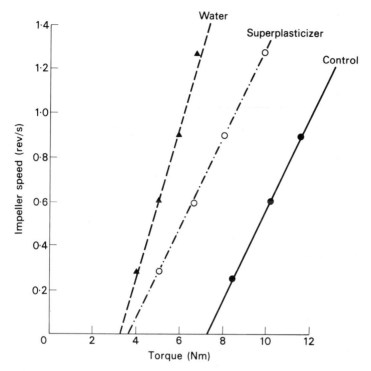

Figure 10.5 Two concretes of the same slump: relative effects of superplasticizer and water.

Mix	Slump (mm)	g (Nm)	h (Nm s)
C	75	7.3	4.8
SP	180	3.6	4.9
W	188	3.2	2.9

Table 10.2 Effect of superplasticizer on mixes of different cement content (Bloomer) – MH apparatus

Cement (kg/m^3)	Water (kg/m^3)	Aggregate* (kg/m^3)	Before addition		After addition		Change (%)	
			g	h	g	h	g	h
240	192	1920	3.0	1.8	1.9	1.9	−37	+5
365	182	1825	4.7	1.2	2.2	1.5	−53	+25
435	195	1740	4.5	2.0	1.4	2.4	−69	+20

*20 mm : Zone 3 sand = 60 : 40.

on three mixes. The slump value of the control mix may be increased either by adding plasticizer or water; the values of g for the two modified mixes are the same, as are the slump values, but the values of h are quite different. The effect of this would readily be observed in practice.

Unfortunately, the effects of adding a plasticizer do not depend only on the amount added because other factors can influence the opportunities for the material to be adsorbed on particles of the cement. A consequence is that in practice concrete containing a plasticizer may show more variation in workability than expected. Al-Shakhshir[12] investigated the effects of a lignosulphonate plasticizer in mixes containing dried and pre-soaked gravel coarse aggregate, and ordinary Portland and rapid-hardening cements from the same clinker, and he used a variety of mixing procedures. He found that, provided the mixing programme was consistent, reproducibility was good for both unplasticized and plasticized mixes, and just as good for the latter as for the former. The values of g were greater by a small but statistically significant amount for the mixes containing rapid-hardening cement than for those containing ordinary Portland, an effect which might be due to either the higher specific surface or the higher sulphate content of the former. It made no difference whether the admixture was added in the mixing water or separately at the same time as the mixing water, but the time at which the addition was made had an important influence on g. If addition was made at the same time as the mixing water, g was increased by 70% above the value obtained if addition was after 1 min of mixing.

Penttala[13], using the slump and Vebe tests, obtained a similar result for both melamine formaldehyde and naphthalene formaldehyde sulphonates. In addition to finding that delayed addition resulted in increased workability, he also found that it resulted in decreased air content.

The explanation is seen in terms of the adsorption of the superplasticizer molecules on to the tricalcium aluminate, which occurs in substantial amounts, even in a few seconds. If the superplasticizer is added with the mixing water, much of it becomes rigidly attached to the unreacted C_3A so that only a small amount remains available for dispersion of the silicate phases, but if addition is delayed, the C_3A has time to develop a protective layer of ettringite so adsorption is reduced, more superplasticizer molecules do remain available, and the plasticizing action is therefore more effective.

On a longer timescale, Bloomer[11] compared the effects of adding a superplasticizer at 15 and 30 min after mixing and found that the reduction in g for the latter was about 10–20% less.

Of course, the presence of a plasticizer or superplasticizer does not prevent the loss of workability with time and in the case of the latter, which is used for very high-workability concretes, the loss may be much greater than is acceptable. It can be ameliorated by delaying the addition, repeated dosing, or mixing a retarder with the super-plasticizer.

Bloomer[11], Rixom and Waddicor[10] and Banfill[14] found that the rates of change of g and h with time depend on both the admixture and the cement type. The rates for both g and h when a melamine resin type is used may be two or three times greater than for naphthalene resins, and for the latter are no greater than for high-workability concrete containing no admixture. Banfill also found that the original flowing consistency of a high-workability concrete could be regained by adding a second dose of admixture after up to 60 min for a melamine resin and up to 120 min for a naphthalene resin.

Fukuda, Mizunama, Izumi, Izuka and Hisaka[15] consider that the loss in workability with time, referred to as **slump loss**, is caused by both physical and chemical factors, the former being an increase in the number of cement particles per unit volume as a result of dispersion, and the latter being the gradual consumption of the dispersant during the cement hydration reaction. They argue that the problem can be solved by incorporating in the superplasticizer a reactive polymer which initially is insoluble but is converted to a soluble dispersant by reaction with the alkali that is always present. Thus the process proceeds in four steps:

(a)   generation of alkali by the hydration of the cement;
(b)   alkali attack on the reactive polymer;
(c)   conversion of the polymer into soluble dispersant;
(d)   adsorption of the dispersant onto the surface of cement particles.

Put simply, the dispersant that is being removed is being replaced by new dispersant.

Fukuda *et al.* present the results shown in Figure 10.6. A mix that did not contain the reactive polymer suffered a decrease in slump from about 180 mm to about 130 mm in 60 min whereas those that did contain it maintained a slump constant at 180 ± 20 mm.

The efficacy of a plasticizer may also be affected by the length of mixing time and some recent results from a site where the time varied from 1 to 5 min are shown in Figure 10.7 for a mix with a lignosul-phonate plasticizer. The value of g decreased as mixing time increased while that of h seemed to increase, although for this small number of results the trends are not statistically significant.

Anomalous behaviour, such as abnormally rapid stiffening of the mix, may sometimes occur, and Coulon[16] and Sugi, Kameshina,

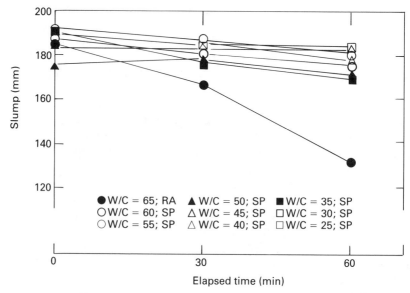

Notes: SP; New superplasticizer containing reactive polymer
RA; Water reducing admixture (lignosulfonate)

Figure 10.6 Effect of incoporation of reactive polymer on slump loss.
(*Fukuda et al.*)

Yamada and Okada[17] have explained cases of this in terms of the relative availability of sulphate and tricalcium aluminate (C_3A) in providing sites for adsorption, and for reaction with each other, or with water.

Edmeades, Hewlett and Thomas reported[18] problems on two separate sites caused by rapid loss of workability of concretes in which sulphate-resisting cement, from the same source in both cases, was being used in a mix that also contained a lignosulphonate-based plasticizer. In each case the start of the trouble coincided with the delivery of a new batch of cement and when that cement was replaced by SRPC from another source the trouble disappeared. No samples of the suspect cement were available and attempts to reproduce the anomalous effect using a fresh sample from the same source were unsuccessful until the cement was conditioned at 150°C for 24 hours. This suggested that the problem batch had been overheated in grinding with resulting partial dehydration of gypsum. They were then able to propose an explanation which assumed supersaturation of the mixing water with respect to gypsum followed by competition for active C_3A and C_4AF sites between calcium sulphate and the organic admixture. The undesirable behaviour of the overheated cements could be prevented by delaying the addition of admixture to the concrete by several minutes

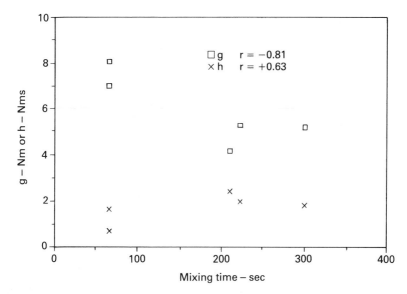

Figure 10.7 Effect of time of mixing on efficiency of a plasticizer.
× g $r = -0.81$
⊙ r $r = +0.63$

so that the normal hydration reactions become established before admixture molecules are available to block reactive sites at which ettringite can form. The practical effect of doing this is shown in Figure 10.8.

The behaviour of an admixture can also be affected by changes in other constituents of a mix such as, for example, the delivery of a pfa with a higher loss on ignition than normal.

For practical purposes the various complications in the use of plasticizers and superplasticizers should be catered for by making a trial mix with an admixture concentration recommended by the manufacturer and then standardizing not only the quantity to be added but also the method, time, and rate of addition and the total time of mixing. Records should be kept of materials deliveries so that if anomalous behaviour occurs there is a possibility of identifying the cause.

10.2 AIR ENTRAINMENT

The commonest use of air entrainment, for improvement of frost resistance of the hardened material, is in pavement-quality concrete for which a typical mix might be Grade 30 with 35 mm slump and about 5% entrained air. Figure 10.9 shows the effect of air content on g and

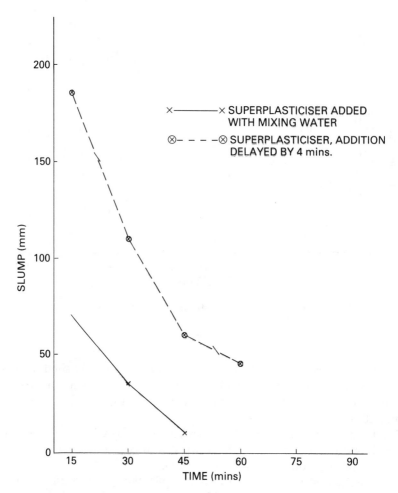

Figure 10.8 Slump loss with time of SRPC concretes containing a superplasticizer. (*Edmeades et al.*)

h for three mixes of this type, whose details are given in Table 10.3. Five per cent air reduces h by about 70% but g by only 30%; there is no further decrease in h beyond an air content of about 5% but g continues to decrease slowly.

10.3 COMBINED EFFECTS

The fact that air-entraining agents affect h more than g, whereas plasticizers affect g more than h, is one that can be exploited in a consid-

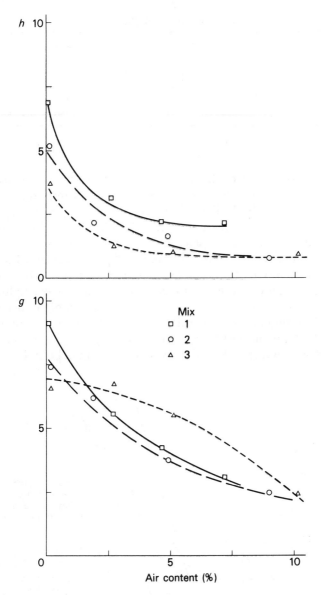

Figure 10.9 Effect of air content on g and h. (LM apparatus; mixes as Table 10.2)

Table 10.3 Details of air-entrained mixes of Figure 10.9

Mix	Cement (kg/m³)	Water (kg/m³)	Aggregate (kg/m³)	Sand (%)
1	300	165	1963	35
2	292	175	1952	40
3	277	180	1956	45

eration of quality control. It can also be used to adjust mix properties by incorporating both types of admixture, and an important practical example was given by Helland[19], in relation to a slip-forming process. Helland's results are shown in Figure 10.10. Mix E, which contains no admixture, behaved satisfactorily but mix F, which is the corresponding 'water-reduced' mix containing a plasticizer and having the same slump as mix E, did not respond well to vibration. This illustrates in a very practical way the point made earlier that a water-reduced mix does not in fact have the same workability as the original, because the value of h is higher. In Helland's case, the h value was reduced by incorporating an air-entraining agent, as well as the plasticizer, to produce mix G which behaved as well as the original mix E.

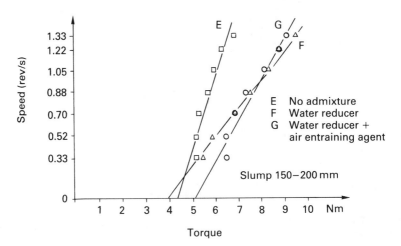

Figure 10.10 Effect of combination of plasticizer and air-entraining agent. (*Helland*)
E   no admixture
F   water reducer
G   water reducer and air-entraining agent

10.4 REFERENCES

1. Hewlett, P.C. (1988) *Cement Admixtures. Uses and Applications*, London, Longman Scientific.
2. Rixom, M.R. (Editor) (1977) *Concrete Admixtures: Use and applications*, Lancaster, Construction Press.
3. Rixom, M.R. and Mailvaganam, N.P. (1988) *Chemical Admixtures for Concrete* 2nd Edition, London, E. & F.N. Spon.
4. Kreijger, P.C. (1980) Plasticizing and dispersive admixtures, *Proceedings of Admixtures Congress CI 80, Lancaster, 1980*, The Concrete Society, Construction Press, 1–16.
5. Hewlett, P.C. (Editor). Superplasticising admixtures in concrete. Publication 45.030. Slough, Cement & Concrete Association, 1976.
6. Scullion, T. (1975) The measurement of the workability of fresh concrete, MA Thesis, University of Sheffield.
7. Waddicor, M.J. Private communication to P.F.G. Banfill, 3 Nov. 1980.
8. Edmeades, R.M. Private communication to P.F.G. Banfill, 7 Jan. 1981.
9. Banfill, P.F.G. (1983) In Tattersall, G.H. and Banfill P.F.G., *Rheology of Fresh Concrete*. Pitman, London, Chapter 13.
10. Rixom, M.R. and Waddicor, M.J. (1981) The role of lignosulphonates as superplasticizers, in *Developments in the use of superplasticizers*, Publication SP-68, Detroit, American Concrete Institute, 359–80.
11. Bloomer, S.J. (1979) Further development of the two-point test for the measurement of the workability of concrete, PhD Thesis, University of Sheffield.
12. Al-Shakhshir, A.T. (1988) Workability of plasticized concrete, Dissertation submitted in part fulfilment of requirements for Degree of M Sc(Eng), University of Sheffield.
13. Penttala, V. (1990) Possibilities of increasing the workability time of high strength concretes, in *Proceedings of RILEM Colloquium on Properties of Fresh Concrete, University of Hannover 3–5 Oct. 1990*, London, Chapman & Hall, 92–100.
14. Banfill, P.F.G. (1980) Workability of flowing concrete, *Magazine of Concrete Research*, **32**(110), 17–27.
15. Fukuda, M., Mizunuma, T., Izumi, T., Izuka, M. and Hisaka, A. (1990) Slump control and properties of concrete with a new superplasticizer. I. Laboratory studies and test methods, in *Proceedings of International RILEM Symposium on Admixtures for Concrete, Improvement of Properties, Barcelona, 14–17 May 1990*, London, Chapman & Hall, 10–5.
16. Coulon, C. (1979) Rheological anomalies introduced by plasticizers based on lignosulphonates, *Silicates Industriels*. **44**, 235–9.
17. Sugi, T., Kameshina, N., Yamada, S. and Okada, K. (1974) Effect of additives on the dissolution of $CaSO_4$ in cement, *Semento Gijutsu Nempo*, **28**, 87–90.
18. Edmeades, R.M., Hewlett, P.C. and Thomas, R.E.R. (1984) An explanation of some anomalous admixture behaviour. Paper presented at Discussion on Research Aspects of Admixtures for Cement, Mortars and Concrete, Building Research Establishment, Garston, 7 June 1984 and also at Concrete Materials Research Seminar, 9–10 July 1984.
19. Helland, S. (1982) Slipforming of concrete with low water content. Paper presented at Annual Conference of Norwegian Concrete Society, Oct. 1982 and at VIIth International Melment Symposium, 1983. Also *Concrete*, **18**, Dec. 1984, 19–21.

11 Effect of cement replacements and of fibres

11.1 INTRODUCTION

There are several materials that can be, and are, used as partial replacements for cement in a concrete mix and since they are produced as by-products in other industrial processes the cynic may think that this is just a convenient way of getting rid of unwanted waste products that would otherwise have no commercial value. In fact there are considerable benefits to be obtained because of the effects they may have on the properties of both the fresh and the hardened concrete. They fall very roughly into two categories, those that consist of more or less pure silica and depend for their effectiveness on their ability to react with the free lime in the mix, and those that contain compounds somewhat similar chemically to those in Portland cement and under the influence of an initiator will hydrate to produce hydration products similar to those produced in the hydration of cement itself. In the first category are pulverized fuel ash (pfa) which is also sometimes referred to as fly ash, microsilica which is also referred to as silica fume, and, in appropriate parts of the world, material such as rice-husk residues. In the second category are ground granulated blast-furnace slag (ggbs) and, possibly in the future, gasifier slag which is obtained in the process of gasification of coal. The ones important at present are pfa and slag, which are in fairly common use, and micro-silica which is relatively new but is likely to be used in increasing quantities.

/The effect on fresh concrete properties of incorporating these materials depends mainly on their particle shape and particle-size distribution, while the effect on hardened concrete properties depends on the way the fresh concrete is affected and also on the chemical properties that influence development of strength and durability. To some extent, chemical properties may also affect the fresh concrete.)

The relationship of these materials to each other can be shown on a ternary $CaO-Al_2O_3-SiO_2$ composition diagram but can also be shown

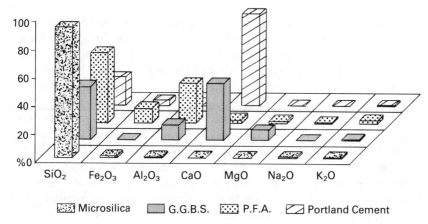

Figure 11.1 Constituents of cementitious materials.

as in Figure 11.1 which is due to Male[1] as is Figure 11.2 which shows the main physical properties.

11.2 PULVERIZED FUEL ASH

Pulverized fuel ash (pfa), which is also known as fly ash, is a by-product of the combustion of pulverized coal in power stations, being removed from the flue gases by mechanical or electrostatic collectors. It consists mainly of spherical glassy particles ranging from 1 to 150 μm in diameter, of which the bulk passes a 45 μm sieve. The finely divided

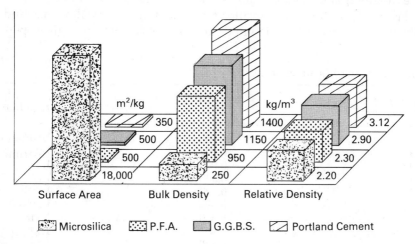

Figure 11.2 Physical properties of cementitious materials.

glassy material consists chiefly of silica with some alumina and will react with the lime that is present in a mix to produce the same calcium silicate hydrate as is produced in Portland cement. When some of the cement in a mix is replaced by pfa early strength is lowered, because there is less cement, but once sufficient lime has been liberated to start the pozzolanic reaction (i.e. the reaction with the silica) the strength increases more rapidly so that eventually the line for strength development as a function of age crosses that for a normal mix without pfa. The main practical advantages are in a lower rate of heat evolution, improved resistance of the concrete to chemical attack, and reduced susceptibility to alkali-silica reaction. In addition, there is a marked effect on the workability of the fresh concrete, caused primarily by the spherical shape of the pfa particles.

The potential improvement in workability can be exploited by reducing the water content sufficiently to counter what would otherwise be a loss in early strength. Table 11.1 gives factors which when applied to an existing satisfactory concrete mix are intended to produce a concrete of the same workability and the same 28-day compressive strength with increased strengths at later ages.

However, Brown[2] found that the effect of pfa replacement on workability depended not only on the actual mix proportions but also on the choice of test used to assess the workability. For compacting factor and Vebe each 10% weight fraction of cement replaced by pfa had approximately the same effect as increasing the water content of the mix by about 4% whereas for the slump test the corresponding figure was about 6%. This is a clear indication that the effects on g and h are different, and work by Ellis[3] shows that this is indeed the case.

Figure 11.3 shows his results for a medium-workability mix and illustrates the fact that it is important to specify whether a given per-

Table 11.1 Mix adjustment factors

pfa in cement (% by mass)	Water content	Cement + pfa content
0	1.000	1.000
15	0.970	1.035
20	0.965	1.050
25	0.945	1.065
30	0.920	1.080
35	0.895	1.095
40	0.870	1.110
45	0.845	1.125
50	0.820	1.140

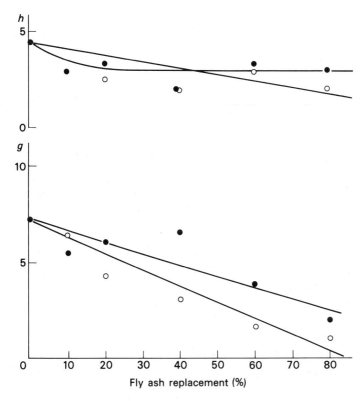

Figure 11.3 Effect on g and h of partial replacement, by mass (●) and volume (○), of cement of fly ash. (LM apparatus; mix 378 kg/m³ cement, 169 kg/m³ water, 1881 kg/m³ aggregate as Zone 3:20 mm = 33:67.) (*Ellis*)

centage replacement is reckoned on a mass or volume basis. The different relative densities of cement and pfa result in an increase in the volume of total cementitious material when replacement is on a mass basis, and this may give a greater reduction in g and h by supplying additional lubricant round the grains of aggregate, or it may give a lesser reduction because of the increased surface area to be wetted in the paste. For these results, the latter appears to be the case because the effect on g and h of volume replacement is appreciably greater than that of mass replacement.

Figure 11.4 shows the effect of replacement by mass for mixes of similar workability initially but of different cement contents and fines contents. The reduction in g is less marked at higher cement contents, presumably because the increase in volume of the cementitious component is more useful in a relatively lean mix, in filling voids and in lubricating. Except in one mix, h is largely unaffected.

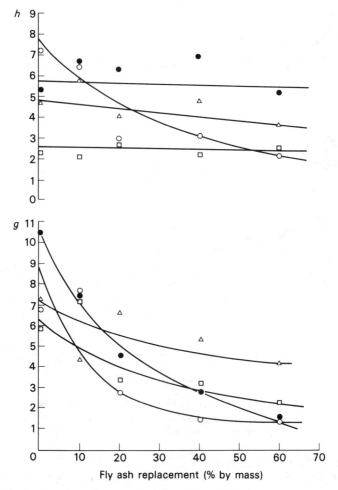

Figure 11.4 Effect on *g* and *h* of partial replacement of cement by fly ash in various mixes (*LM apparatus*).

Banfill[4] has used these results, together with the known information about the effect of change in water content on *g* and *h*, to estimate water-reduction factors for comparison with those given in Table 11.1, and his results are shown in Figure 11.5. Two factors stand out. First, the water adjustment to achieve equal *g* values is much larger than that needed to achieve equal *h* values; in other words, it is impossible to apply a water adjustment to obtain a pfa-replaced concrete whose *g* and *h* values are identical with those of the plain concrete. If the larger reduction, appropriate to give constant *g* is used, then *h* will be too

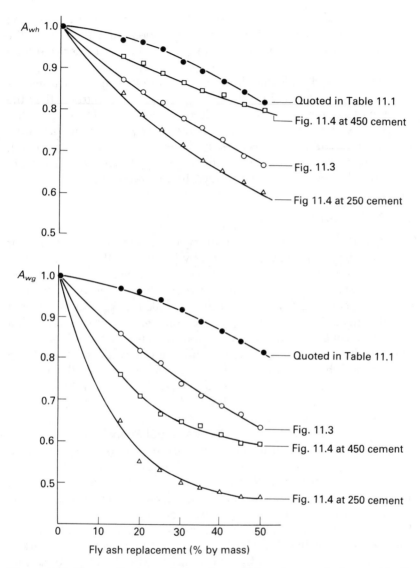

Figure 11.5 Water-adjustment factors for partial replacement of cement by fly ash. (*Banfill*)

large; conversely, if the smaller reduction, appropriate for constant h is used, g will be too small. The situation is thus rather similar to that when water reduction is used in conjunction with addition of a plasticizer, discussed previously. Brown's finding of the effect of test method, referred to above, is also explained because the compacting factor and Vebe tests operate at a higher shear rate than does the slump test, so the effect of changes in g is less in the former two than in the last.

Secondly, Banfill's results show that required water reductions are larger than those given in Table 11.1. It may be remarked that the reductions recommended in the table are intended to be used with a simultaneous increase in the total cementitious content, whereas in the mixes used by Ellis on whose results Banfill's calculations are based, no such increase was made. Banfill considered whether this could explain the discrepancy and concluded that it could not.

11.3 MICROSILICA

Microsilica (MS), which is also known as condensed silica fume, or sometimes simply as silica fume, is a by-product in the production of silicon and ferrosilicon alloys and is collected from the gases issuing from the furnace. It consists of microspheres of very pure glassy silica and has a specific surface of the order of $18,000 \, m^2/kg$ which makes it an ideal material to take part in a pozzolanic reaction. It is claimed that a microsilica concrete can achieve without difficulty high early strength to the extent that $40 \, N/mm^2$ that would normally be reached in 28 days, can be achieved in about 30 hours and increase to $80 \, N/mm^2$ at 28 days. The resulting concrete is said to be highly resistant to abrasion, resistant to chemical attack, and immune to alkali-silica reaction[1] provided the MS is fully dispersed.

In its initial form as a powder, MS has a bulk density of only about $250 \, kg/m^3$, so it is difficult to handle and expensive to transport. It can also be difficult to disperse properly so for these reasons it is often prepared in the form of a slurry consisting of 50% MS by weight in water, and is used in that form. A concrete mix may contain of the order of 16 to 25% slurry by weight of cement.

Since incorporation of pfa has such marked effects on workability it would be expected that MS would have too, to an even more marked extent, and such is the case. Mixes containing MS are cohesive, so that they can be very effective for underwater concreting, and in more ordinary use they do not bleed. Male[1] emphasizes that the usual slump test is useless and misleading if applied to MS concrete, and says that it is possible to get a negative slump from a well-designed mix with well-dispersed microsilica when the concrete sticks to the cone as it is lifted. In practice, the aim may be to produce a concrete with a workability 'equivalent' to a 75 mm slump normal concrete, and as this may often mean that the slump of the MS concrete is about 15

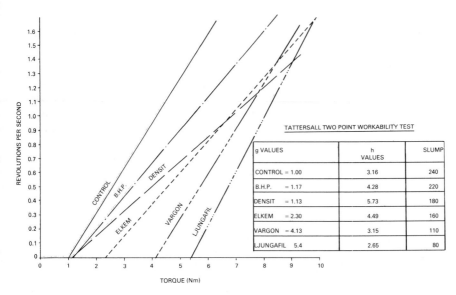

Figure 11.6 Effect of source of microsilica.(*Munn*)

to 20 mm, it may be necessary to make the judgement largely on a subjective basis to achieve a workability suitable for site. Complaints have sometimes been made that an MS concrete would not pump, when in fact the difficulty was that the material had refused to pass through the grid of the pump hopper. All that was necessary was to provide extra energy by dropping the material a little further or to apply mild vibration to overcome the yield value. The concrete then pumped easily. In fact a small addition of microsilica, say about 2 to 3% solids by weight of cement, will act as a pumping aid for normal concrete, and is so effective that a 150 kg/m^3 mix with a small addition of microsilica will pump satisfactorily.

In microsilica concrete, the effect of a change in sand content is greater than that in normal concretes, and would require corespondingly a greater change in water content to maintain workability. Usually, microsilica concrete also contains a plasticizer such as a refined lignosulphonate to ensure adequate dispersion of the MS.

Although the effects on workability are so profound, very little has so far been published on the results of proper measurements. Munn[5] compared microsilica from five different sources by using each of them as a partial replacement ($7\frac{1}{2}$%) for the cement in a mix of 10 mm maximum size, aggregate:cement ratio 3.9:1, and water/cement ratio 0.34. His results are shown in Figure 11.6 and he concludes that the two-point test is readily capable of differentiating between the different microsilicas, but the possibility of effects due to differences in dispersion can not be discounted.

According to Wallevik[6] substitution of cement with silica fume up to a threshold value, which depends on cement content, reduces plastic viscosity h by up to 50%. The threshold value is thought to decrease with decreasing water content. Yield value, g, is nearly constant until the threshold value is reached and then increases dramatically. This is shown in Figure 11.7.

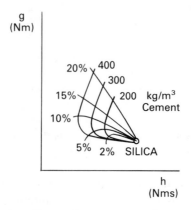

Figure 11.7 Effect of microsilica replacement at various cement contents. (*Wallevik*)

11.4 BLAST FURNACE SLAG

Blast furnace slag is a by-product of the production of iron which, when cooled rapidly in intense water sprays, solidifies in a glassy form that has latent cementitious properties. After cooling, the material, which is in a granular form, is either ground with Portland cement clinker and gypsum, or is ground alone to produce ground granulated blast-furnace slag (ggbs) which is then mixed with Portland cement at the point of use. Slag normally requires an alkaline medium to initiate its hydration and in Portland blast-furnace cement, prepared by either of the two methods, that is provided by the gypsum and the lime liberated when the Portland component hydrates. The advantages of incorporating ggbs lie in a lower initial rate of heat evolution and in an enhanced resistance of the concrete to chemical attack.

The effect on workability of the fresh concrete of replacing part of the Portland cement with ggbs is much less than that of a similar replacement by pfa. Early work with ggbs of specific surface around

$350 \, m^2/kg$ suggested the effect was probably equivalent to increasing water content by about 5%.

In normal practice where ggbs is used the aim would be to work to a constant slump with water contents reduced if possible. An investigation was conducted on these lines with a mix of aggregate:cement ratio 6.4:1 and 38% fines with cement/ggbs replacements of 0, 10, 30, 50, 70 and 90% W/W. The water content was adjusted in each case to give a constant slump of 75 mm ten minutes after mixing, two mixes were made at each replacement level, and the order of making and testing was randomized. Water content of the aggregates (irregular gravel and Zone 2 sand) was measured for each individual batch and appropriate allowances were made. Workability was assessed by the two-point method in the LM (planetary) mode at 20 min and 60 min and also slump was measured a second time at 60 min.

The results are shown in Figure 11.8 and confirm the observation that for constant slump the water-content reduction possible does not exceed about 5%. In no case is the value of g and h reduced below that for the control mix. The value of g is significantly higher than the control at 30 and 50% replacement, while h is significantly higher at 70 and 90% replacement, and the differences tended to increase with time. There is no ready explanation of the shape of these curves, but even though the effects are significant they are not large and the general statement that ggbs has little effect seems to be confirmed.

However, recent work by Ellis and Wimpenny[7] shows that the statement requires some qualification, particularly for mixes richer or leaner than an aggregate:cement ratio of 6:1. In a thorough investigation they studied slags from two sources (type A and type B) at three replacement levels (0, 40 and 70%) in concretes of three different cement contents (200, 300, and 400 kg/m^3). Cement content here refers to total cementitious content, i.e. cement plus ggbs. The aggregate: cement ratios corresponding to the three cement contents were respectively 10.0:1, 6.4:1, and 4.6:1, with water/cement ratios respectively of 0.83, 0.55 and 0.41. The volume of cement plus fine aggregate was kept approximately constant in each of their 15 mix designs.

Workability was assessed 15 min after the first addition of mixing water by means of the two-point test (LM mode, planetary motion, nine experimental points), and slump and compacting factor were also measured. Each mix design formed part of a randomized block of 15 mixes and the whole experiment was repeated three times to give four such randomized blocks on which an analysis of variance was carried out for each of the measured parameters, g, h, slump and compacting factor.

Their results are shown in Tables 11.2 & 11.3 and in Figures 11.9 &

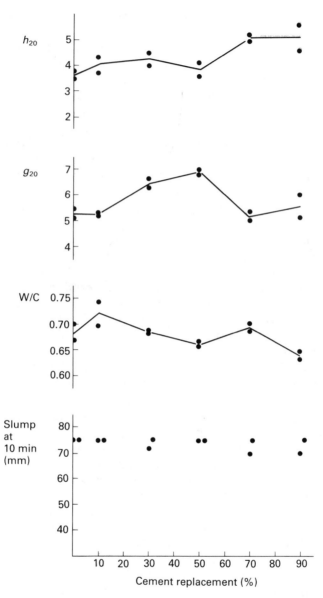

Figure 11.8 Effect of ggbs replacement on mixes of constant slump (suffix denotes time [min] between mixing and measurement).

Table 11.2 Workability properties of fresh concrete: two-point test results, 'g' (Ellis and Wimpenny)

	Slag type A								
Cement content	$200 \, kg/m^3$			$300 \, kg/m^3$			$400 \, kg/m^3$		
Replacement (%)	0	40	70	0	40	70	0	40	70
Block 1	6.3	2.5	0.8	2.8	2.0	1.4	2.4	3.1	2.6
Block 2	6.4	4.4	0.5	3.0	1.7	1.6	3.5	3.7	2.6
Block 3	6.2	2.6	1.8	2.8	2.1	1.2	2.3	2.4	3.2
Block 4	4.0	1.7	0.5	1.7	1.9	1.4	2.1	2.3	2.1

	Slag type B								
Cement content	$200 \, kg/m^3$			$300 \, kg/m^3$			$400 \, kg/m^3$		
Replacement (%)	0	40	70	0	40	70	0	40	70
Block 1	6.3	4.9	4.6	2.8	3.9	5.4	2.4	5.2	7.5
Block 2	6.4	5.3	1.5	3.0	2.1	2.6	3.5	5.3	6.6
Block 3	6.2	1.6	1.2	2.8	2.5	4.7	2.3	5.8	6.1
Block 4	4.0	4.1	1.1	1.7	2.8	2.6	2.1	3.7	6.7

Table 11.3 Workability properties of fresh concrete: two-point test results. 'h' (*Ellis and Wimpenny*)

	Slage type A								
Cement content	$200 \, kg/m^3$			$300 \, kg/m^3$			$400 \, kg/m^3$		
Replacement (%)	0	40	70	0	40	70	0	40	70
Block 1	5.5	7.8	7.3	3.0	3.3	2.5	2.5	3.1	3.0
Block 2	4.8	7.5	6.8	3.3	3.3	3.0	2.8	2.8	3.6
Block 3	4.9	5.9	7.5	4.3	4.5	3.6	3.6	2.8	3.8
Block 4	5.2	6.1	6.2	3.1	3.0	3.0	2.8	2.7	2.9

	Slage type B								
Cement content	$200 \, kg/m^3$			$300 \, kg/m^3$			$400 \, kg/m^3$		
Replacement (%)	0	40	70	0	40	70	0	40	70
Block 1	5.5	4.4	7.1	3.0	3.4	2.8	2.5	2.7	5.1
Block 2	4.8	6.0	4.9	3.3	3.3	3.6	2.8	3.8	4.6
Block 3	4.9	8.0	7.4	4.3	4.1	4.8	3.6	3.4	4.3
Block 4	5.2	4.9	5.6	3.1	3.1	3.9	2.8	3.4	4.6

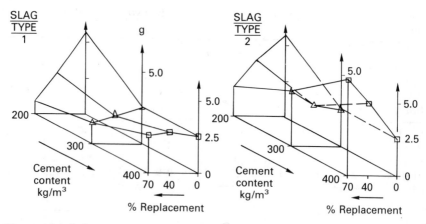

Figure 11.9 Influence of cement content (C) and replacement level (R) upon two-point test parameter *g* for OPC/ggbs concretes containing slag type A and slag type B. (*Ellis and Wimpenny*)

11.10. In these figures the results are plotted on a quasi-three-dimensional basis using an oblique parallel projection, so that plots of the dependent variable (vertical axis) versus replacement level (horizontal axis) are placed at equal intervals along an oblique line representing cement content.

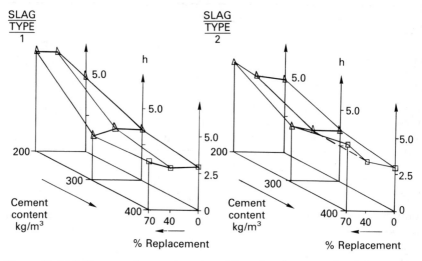

Figure 11.10 Influence of cement content (C) and replacement level (R) upon two-point test parameter *h* for OPC/ggbs concretes containing slag type A and slag type B. (*Ellis and Wimpenny*)

It is clear that for both g and h the effect of replacement level R depends on the cement content C, that is, there are interactions between the variables, and this is confirmed by the analysis of variance. For the intermediate cement content of $300\,kg/m^3$, which corresponds to an aggregate:cement ratio of 6.4:1, replacement of OPC by ggbs has little effect on either g or h. Looking at the results more broadly shows that for slag A, g decreases substantially with increase in R for $C = 200\,kg/m^3$ but the dependence on R decreases as C increases until at $C = 400\,kg/m^3$ R has no effect. For slag type B this trend is more marked in that a decrease in g with increase in R at $C = 200\,kg/m^3$ changes to an increase in g with increase in R at $C = 400\,kg/m^3$. The behaviour of h is less dependent on slag type. The effect of change in replacement level R at constant cement content C is not large but on the whole h increases slightly as R increases. At constant R, h decreases appreciably as C increases, this being more marked for slag A. In general, the slump and compacting factor results show trends similar to those for g, although of course, in the opposite direction numerically.

Ellis and Wimpenny conclude that the factors cement content, replacement level, and slag type, and their interactions, all have an effect on workability, and they point out that it is not possible to assess that effect properly without a knowledge of the results from the two-point test because g and h may change in opposite directions.

Finally, it may be noted that the two slags used came from two different sources. They differed from the slag used in the experiments described earlier in being finer, with specific surfaces of 406 and $452\,m^2/kg$ respectively and having $45\,\mu m$ sieve residues of 18% and zero.

11.5 FIBRES

The practice of incorporating fibres as reinforcement in ceramic materials is a very ancient one[8] but its introduction into concrete technology is relatively recent. The object is to improve toughness, impact resistance, and strain capacity, and to reduce the tendency to crack propagation. The fibres may be introduced by means of hand laying-up or by a spraying technique, or they may be included in a mix that is mixed and placed more or less conventionally. It is only this last case that is considered here because in the first two the fibre-reinforced composite does not exist other than in its final moulded form, and consequently the flow properties of the fresh material are unimportant except as they influence deformation under self-weight.

The fibres used in concrete are of polypropylene or steel; the former is available in fibrillated form and the latter in various configurations that have been developed in attempts to improve the mechanical bond between the fibre and the matrix. In considering their effect on work-ability, fibres may first of all be thought of as an aggregate with extreme deviation in shape, with the additional property that they are flexible, so it is not surprising that their inclusion in a mix decreases work-ability. There are difficulties in achieving uniform dispersion which increase as fibre length increases, and tangling and balling can occur, particularly for the longer fibres. The size and concentration of aggregate has an important influence on the effect of the fibres. Hannant[9] recommends that the coarse aggregate content be lower than in a normal mix and that the maximum particle size should not exceed 10 mm. The mortar volume (all particles below 5 mm) should be around 70% and aggregate:cement ratios as low as 3:1 are needed.

The fibre factors that are of the greatest importance are the aspect ratio (length/diameter, l/d) and the fibre volume concentration.

A new test for workability of fibre-reinforced concretes has been proposed[10] but has been severely criticized[11]; the single-point test that has been most commonly used is the Vebe test.

Edgington, Hannant and Williams[12] carried out experiments in which steel fibres of a variety of shapes and surface conditions, of diameters from 0.15 to 0.50 mm and aspect ratios from 30 to 250, were incorporated at various concentrations into cement paste, mortar, 10 mm concrete, and 20 mm concrete. They measured both Vebe time

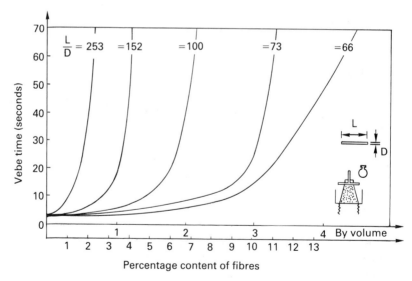

Figure 11.11 Effect of fibre aspect ratio on Vebe time of fibre-reinforced mortar. (*Edgington, Hannant and Williams*)

and compacting factor but considered that results for the latter were suspect because all the mixes required rodding through the hoppers. They present the Vebe results shown in Figure 11.11 as typical for fibre-reinforced mortars and emphasize that it is not the actual length or diameter of fibre that is important but the ratio of the two, that is, the aspect ratio. It is clear that for a given Vebe time a higher volume of low-aspect-ratio fibres may be incorporated in a mix, or conversely, the higher the aspect ratio the lower the volume concentration that can be used.

Figure 11.11 also indicates that there is a critical fibre content beyond which Vebe time increases very rapidly. Figure 11.12 has been constructed from the results of Edgington *et al.* by assuming that the critical fibre content may be defined as the fibre content at the point of intersection of the tangents drawn to the initial and final parts of the curves shown in Figure 11.11, and it shows how critical fibre content depends on the aspect ratio of the fibres.

Figure 11.13, also due to Edgington, Hannant and Williams, shows the dependence of Vebe time on fibre content for fibres with an aspect

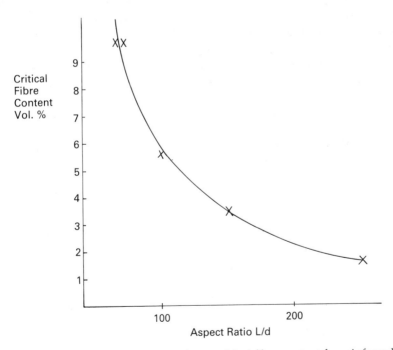

Figure 11.12 Effect of fibre aspect ratio on critical fibre content for reinforced mortar. (Calculated from results of *Edgington, Hannant and Williams*)

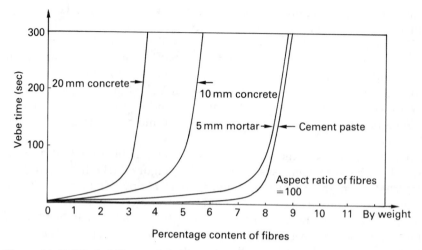

Figure 11.13 Workability against fibre content for matrices with different maximum aggregate size. (*Edgington, Hannant and Williams*)

ratio of 100 in matrices ranging from cement paste to concretes with 20 mm aggregate. The form of the relationship, already shown for mortars, is similar for other matrices and the critical fibre content decreases as the size of the coarse aggregate increases. The characteristics of the fibre-reinforced paste and mortar are very similar, thus indicating that the presence of aggregate particles up to 5 mm in size has little influence.

According to Edgington *et al.*, a reasonable estimate of the fibre content required to make concrete effectively unworkable can be obtained from the following equation:

$$CFC = 75 \cdot \frac{\pi SG_f}{SG_c} \cdot \frac{d}{L} \cdot \frac{W_m}{W_m + W_a} \qquad (11.1)$$

where

CFC = critical percentage of fibres, by weight of matrix
SG_f = specific gravity of fibres
SG_c = specific gravity of concrete matrix
d/L = reciprocal of fibre aspect ratio
W_m = weight of mortar fraction (i.e. particles under 5 mm)
W_a = weight of aggregate fraction (i.e. over 5 mm)

They recommend that to ensure that the workability of a composite is 'sufficient to allow compaction by substantial external vibration' the

Table 11.4 Behaviour of steel-fibre-reinforced concretes in the two-point apparatus (LM mode)

Fibre content (vol. %)	Fibre size L × d (mm)	0.45	0.50	Water/cement ratio 0.55	0.60	0.65
0.5	20 × 0.3	Too				
1.0	20 × 0.3	stiff-		Tested satisfactorily		
0.5	25 × 0.4	torques				
1.0	25 × 0.4	too				
0.5	60 × 0.65	high				
1.0	60 × 0.65			Fibres tangled during test		

fibre content in practice should not exceed three quarters of this value of CFC.

They also made a study of the air content of compacted steel-fibre-reinforced specimens because they had noted that it had been reported by Grimer and Ali[13] that addition of glass fibres resulted in considerable air entrapment. They found that, provided the composite was capable of compaction on a vibrating table, the air content of the matrix within fibre-reinforced concretes was no greater than that of the matrix without fibres. In fact, in the case of mortars, there was a trend for decreasing air content with increasing fibre content.

Only limited data are available so far from two-point testing of fibre-containing concretes. Steel fibres of the Duoform indented type were used in three sizes, quoted as L × d in mm, 20 × 0.3, 25 × 0.4, and 60 × 0.5. Each of these types was used at fibre contents of 0.5 and 1.0% by volume (1.7 and 3.3% by weight) in a mix of aggregate: cement ratio 4:1 and water content was progressively increased, with remixing between tests, to give a range of water/cement ratios. The LM (planetary) apparatus was used and Table 11.4 shows the qualitative observations on its ability to cope with the various mixes. At low water/cement ratios the torques were too high to be measured but that difficulty could be overcome simply by altering the gear ratios on the apparatus. Of the mixes that tested satisfactorily, all gave points that conformed to a straight line, about half of them with correlation coefficients greater than 0.99, showing that their flow properties are well represented by the Bingham model, as are those of ordinary plain concretes.

Variation of g and h with water/cement ratio for the different mixes is shown in Figure 11.14, from which trends emerge that are in broad agreement with those noted from single-point tests. Increasing fibre

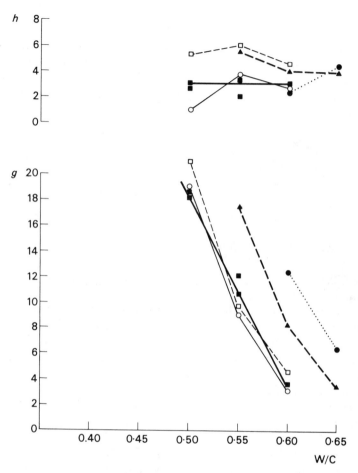

Figure 11.14 Effect of steel fibre on g and h. (LM apparatus; mix A:C = 4.0:1, 10 mm: Zone 3 sand = 50:50.)

Symbol	Fibre content (volume %)	Size (mm)
■	0.5	20
○	0.5	25
●	0.5	60
▲	1.0	20
□	1.0	25

content increases both g and h, whereas increasing length of fibres increases g but has little effect on h.

Only a very few two-point test results have been reported for polypropylene fibres, and these were obtained by Llewellyn[14], using the apparatus in its LM form[15], during the course of an investigation concerned mainly with plastic cracking and the properties of the hardened concrete. He incorporated 40 mm fibrillated polypropylene fibres at three concentrations up to about 0.1% by volume, in mixes containing 20 mm gravel (15) with 42% fines at an aggregate:cement ratio of $6\frac{1}{4}$:1 and water/cement ratios of 0.47 and 0.67. He reports difficulties arising from tangling of the long flexible fibres, during test, but his results are compatible with those for steel fibres in that workability was decreased and the effect on g was greater than that on h.

11.6 REFERENCES

1. Male, P. (1989) An overview of microsilica concrete in the UK, Part 1, Properties of microsilica concrete, *Concrete*, **23**(8), 31–4; Part 2, Applications, *Concrete*, **23**(9), 35–9.
2. Brown, J.H. (1980) The effect of two different pulverized-fuel ashes upon the workability and strength of concrete. Technical Report 536, Slough, Cement & Concrete Association, 18pp.
3. Ellis, C. (1981) Discussion of 'The effect of pulverized fuel ash upon the workability of cement paste and concrete' by D.W. Hobbs, *Magazine of Concrete Research*, **33**(117), 233–5 (Paper: *ibid.*, **32**(113), 219–26; author's reply: *ibid.*, **33**(117), 241.
4. Banfill, P.F.G. in Tattersall, G.H. and Banfill, P.F.G, *Rheology of Fresh Concrete*, Pitman, London, Ch. 14.
5. Munn, C.J. (1986) Compositional variations between different silica fumes, and their effect on early strength in cement and concrete. Project Report in Advanced Concrete Technology Course, Cement & Concrete Association, 41pp.
6. Wallevik, O.H. Private communication.
7. Ellis, C. and Wimpenny, D.E. (1989) A factorial approach to the investigation of concretes containing Portland blast furnace slag cements, in *Proceedings of the 3rd CANMET/ACI International Conference on Fly Ash, Silica Fume, Slag, and Natural Pozzolans in Concrete, 18–23 June, 1989, Trondheim.* Supplementary Papers compiled by M. Asasali; 756–75.
8. Anon., *Exodus*, Ch. 5, vv. 6–19.
9. Hannant, D.J. (1980) Review of the present scene, in *Proceedings of Symposium on Fibrous Concrete, CI 80*. The Concrete Society. Lancaster, Construction Press, 1–15.
10. ACI Committee 544 (1978) Measurement of properties of fibre reinforced concrete. *Proceedings of the American Concrete Institute*, **75**, 283–9.
11. Tattersall, G.H. and Banfill, P.F.G. (1983) *Rheology of Fresh Concrete*, London, Pitman, 237–9.
12. Edgington, J., Hannant, D.H. and Williams, R.I.T. (1974) Steel fibre reinforced concrete, Current Paper CP 69/74, Watford, Building Research Establishment, 17pp.
13. Grimer, F.J. and Ali, M.A. (1969) The strength of cements reinforced with glass fibres, *Magazine of Concrete Research*, **21**(66), 23–30.
14. Llewellyn, D.H. (1990) The effect of polypropylene fibres on the properties of concrete and mortar, B Eng Project, Sheffield City Polytechnic, 95pp.
15. Ellis, C. Private communication, 8 August 1990.

12 Workability and practical processes

12.1 MIXING

A study of the relationship between workability and the practical process of mixing could of course be valuable to mixer manufacturers for application in the processes of mixer design and determination of power requirements, but for the concrete producer and user the interest lies in a consideration of the efficiency of the mixer with which he is provided, and in the possibilities of introducing at the mixing stage a means of control of production. This last will be discussed later.

The question of between-batch variation, that is the variability of nominally identical batches, will also be considered later, in relation to quality control, so discussion here will be confined to within-batch variance, which is a measure of the ability or otherwise of a mixer to produce a well-mixed homogeneous product.

It should perhaps be noted here that firms in the ready-mixed concrete industry operate both 'wet batch' and 'dry batch' plants. In the former case, the batching and mixing plant forms one unit and the mixer itself may be of the pan, tilting drum, or other type; most commonly the operation is of the batch type but in some plants a continuous mixer is used. At dry-batch plants the raw materials are weighed and discharged unmixed into a truck mixer, in which the mixing process takes place. Clearly, the demands on the efficacy of the truck mixer are greater in this case than if it is simply being used as a delivery vehicle for wet-batched concrete.

The degree of uniformity of concrete in a batch depends not only on the design of the mixer but also on the method of loading of aggregates, cement and water. In very bad cases, the concrete may contain dry balls of unmixed material or there may be a delivery of a slurry at one side of the chute with comparatively dry material at the other side, so that even the most casual visual inspection shows that the concrete is unsatisfactory and should be rejected. To avoid this sort of occurrence the *Manual of Quality Systems for Concrete*[1], of the Quality Scheme for Ready Mixed Concrete, requires that 'wherever possible the cement and aggregates should be fed into the drum simultaneously and at a uniform rate', and that the mixer should not be overloaded, but it says nothing about the time and rate of addition of water and admixtures. Very rapid addition of water can result in the obviously heterogeneous concretes referred to above, and haphazard addition of a plasticizer can result in wide variations in workability between nominally identical batches.

However, even when considerable care is taken there may still be considerable within-batch variability of the concrete. This is clearly, although tacitly, recognized in BS 1881: Part 101 when it lays down procedures for sampling, and in the later parts of the same Standard, when requirements for further homogenization of the samples are given. If the mixing process were fully satisfactory there would be no need to state sampling requirements, because two samples from any two parts of the same batch would be sufficiently identical in composition and properties.

Stubbs[2] carried out a thorough investigation on variability in a truck mixer, in terms of cement content as assessed by means of the rapid analysis machine (the RAM). Practical considerations limited his work to one mix only, but he chose a 'popular' mix of $270 \, kg/m^3$ cement content and 75 mm slump. He studied ten nominally identical batches of this mix in one particular truck mixer, and then one batch in each of ten different truck mixers, in each case taking samples when 20, 40, 60, and 80% of the concrete had been discharged, and splitting each sample into two parts for analysis so that an independent estimate of testing error could be obtained. He carried out an analysis of variance with the results shown in Table 12.1.

The interaction term here is significant and there is some justification for carrying out a more detailed analysis than was actually done by Stubbs, but it is perhaps sufficient to point to one or two conclusions. Stubbs found in general that the variance due to sampling was not statistically significant, and somewhat incautiously concluded that

Table 12.1 Stubbs' analysis of variance

Source of variance	Degrees of freedom	Mean Squares
Batch B	9	2750.5
Sampling S	3	208.7
$B \times S$ Interaction	27	355.0
Testing error	40	152.7

therefore all the mixers were satisfactory. The correct conclusion from this finding is that any variability in the mix was such that it could not be convincingly detected by the testing procedure (the RAM) adopted, so it is necessary to consider what is the actual value of the variance. The value quoted above, by Stubbs, is 152.7 which corresponds to a standard deviation of $12.4\,kg/m^3$ and a repeatability, r, (see Chapter 2) of $35\,kg/m^3$. The manufacturers of the RAM claim that a repeatability figure of about half of this, at $20\,kg/m^3$, has been well established by many experiments, and recommend that if repeat determinations in practice differ by more than $20\,kg/m^3$ they should be rejected[3]. Whether that recommendation is regarded as reasonable or not, the fact remains that Stubbs, in a well-conducted practical investigation, found the higher figure. An estimate of Stubbs' sampling variance, that is, the variance in the mixer, may be obtained from the figures for S and $B \times S$ in Table 12.1 as 340 which corresponds to a standard deviation of $18\,kg/m^3$ (this includes variability due to testing). These figures indicate that the variability within a mixer, and therefore of the concrete placed in a job, may be very considerable. One may note also the substantial batch-to-batch variance of nominally identical batches, represented by the figure against B in Table 12.1.

The two-point test has been used in site trials to estimate both between-batch and within-batch variability. One trial, using the MH, or uniaxial, form of the apparatus, was carried out on a site where piles were being cast by tremieing high-workability concrete under water, and measurements were made over a period of some weeks. Over the whole period of the trial some batches were sampled at a quarter, half, and three quarters of the way through the delivery, in order to obtain an estimate of within-batch variability for comparison with the figures for between-batch variability. The results are shown in Table 12.2.

The between-batch variance for h is significantly greater than the within-batch variance at higher than the 0.001 level, but for g, the between-batch variance was significantly greater than the within-batch variance only between the 0.05 and 0.01 levels. In other words, for g, a substantial contribution to the overall variability of the concrete is a

Table 12.2 Variability assessed by workability measurement

Source	Variance		Degrees of freedom
	g	h	
Within-batch	0.358	0.093	14
Between-batch	0.955	0.647	50

variability occurring within the batch. In the trial, the measured values of g ranged from 1.02 to 3.40 and of h from 0.33 to 2.5, an astonishing range for concrete that seemed to be well produced. This is not an isolated example; similar large variability of supposedly identical batches has been found in other site investigations.

Beitzel[4-9] used one particular mix (maximum aggregate size 22 mm, 350 kg/cu.m cement, W/C 0.42, slump 15–50 mm) to investigate the efficiency of a variety of mixers at mixing times of half, one, and three minutes. In each case he followed the manufacturer's instructions for the use of a mixer and sampled into 60 containers that were passed under the discharge on a conveyor belt. Twenty of these samples were selected for determination of water content by drying over a gas flame, and another twenty for determination of fines plus cement content, and coarse aggregate content, by washing through sieves. For each of the variables he calculated the coefficient of variation and concluded that for each mixer there was an optimum mixing time beyond which segregation began. For drum mixers that time was about three minutes, for pan mixers between one and three minutes, and for horizontal shaft mixers between a half and one minute. Of course, since tendency to segregate depends very much on mix composition, these figures may well need modifying for different mixes.

BS 3963:1974 *Testing the performance of concrete mixers* lays down a fairly complicated method '. . . principally for use by the manufacturer (of the mixer) in order to provide evidence that the mixer meets the performance requirements . . .' It involves making three batches of a specified mix, splitting each batch into quarters, and from each quarter taking two independent samples which are each analysed for percentage water, fines, and cement. From the average range D of the results from the twelve pairs of samples, and from the average range C of the results from the four quarters, quantities are calculated which are to be compared with, and to be not greater than, a 'maximum permitted variability' given in a table appropriate to the mixer being considered. The Standard does not make it clear that the calculated figures are in fact estimates of the respective standard deviations, calculated by a rather unsatisfactory method. Neither is it made clear that the permitted variability in a mixer that is deemed to be satisfactory can be rather large. The mixers tested by Stubbs would comfortably comply with the requirements and, for example, the Standard classifies as satisfactory a mixer in which there is a 1 in 10 chance that the fines content of a nominal 40% fines mix could lie outside the range 35 to 45%.

There is little doubt that more work could profitably be done on the design and operation of concrete mixers.

12.2 PUMPING

Pumping as a means of transporting concrete, both horizontally and vertically, is now commonplace (see Figure 12.1), and, when it is used, it imposes further requirements for workability, because the behaviour of the material must be considered not only in relation to placing and compacting but also in relation to its performance in the pumping system.

When a Bingham material is made to flow through a pipe at relatively low velocities it would be expected to exhibit the phenomenon of plug flow, because it is only near the wall of the pipe that the shear stresses are high enough to overcome the yield stress. This means that the concrete flows forward as a solid plug in a way quite different from the way an ordinary liquid would flow under the same conditions, and indeed this can be seen by ordinary observation on site. However, the situation is a little more complicated than at first appears.

From his experimental results for the yield value of a concrete, Morinaga[10] calculated that the radius at which flow would occur, under the conditions he was using, was greater than the radius of his pipe, which means that if the concrete was of uniform composition across the whole cross-sectional area of the pipe, it would not move at all. But flow **did** occur, so some explanation must be sought; it is to be found in the fact that the composition is not uniform across the section and that near the wall there is a layer of cement grout which forms a slippage layer. Later results from the work of Sakuta, Yamani, Kasami and Sakamoto[11] showed that under some conditions flow took place in the main bulk of the concrete as well as in a slippage layer.

Earlier, Aleekseev[12], Weber[13], and Ede[14] had all found that pumping pressure for a given velocity was linearly related to the length of pipe-line and Weber added that, for a given length, pressure was linearly related to velocity of the concrete.

All these results can be explained by a theoretical treatment[15] assuming that one must consider the flow of a Bingham material (the concrete) lubricated by another Bingham material (the slippage layer of grout) of lower yield value and plastic viscosity. These observations are in full conformity with practical observations on the site and it is normal practice to establish a lubricating layer by pumping a grout through the pipe before any attempt is made to pump concrete. Whether the concrete will then continue to pump depends on whether it is capable of sustaining a lubricating layer, and it is its behaviour in this respect, rather than its own bulk properties, that is the determining factor. This statement applies only to straight runs of pipe of unvarying cross-section; when the concrete is forced to flow through a reducing section, or round a bend, the bulk properties will become important in their own right.

There are two extremes of concrete composition that inhibit the use of pumping. At one end of the scale the mix may be so coarse and harsh that it acts as a filter through which water is pumped and lost, while at the other end the mix may be so cohesive that it will not expel grout to sustain the lubricating layer. Between these two extremes lie

Figure 12.1 Placing concrete by pumping and compaction by use of an internal vibrator. (*Photo courtesey of the British Cement Association*)

the pumpable mixes, and to establish which they are it is necessary to consider the bleeding characteristics under pressure, and the void content of the combined coarse and fine aggregates.

Browne and Bamforth[16] emphasize the need for a lubricating layer and draw attention to a statement of Ede's that concrete can be transformed from the unsaturated state (unpumpable) to the saturated state (pumpable) by an increase in the water/cement ratio. They point out that if the permeability of the concrete, under the pressure gradients imposed by the pump, is such that water flows at an excessive rate down the pipeline, then the concrete near the pump may become 'dewatered' and be transformed from the saturated to the unsaturated state, resulting in a large increase in flow resistance and possible blockage.

Browne and Bamforth therefore carried out pressure bleed tests, in which they measured the volume of water, V, expressed from the concrete in time t, when it was subjected to a constant pressure applied by a piston in a specially designed apparatus. Some typical results are shown in Figure 12.2. It appears that the important factor is not the relative positions of the pumpable and unpumpable curves but their shapes, in that the unpumpable concrete tends to have a higher flow rate of water immediately the experiment starts. For all concretes, it was found in practice that the volume of water emitted after the first 140 s was small, so the volume up to that time, V_{140}, could be taken as the total available. Hence, the difference between this quantity and the

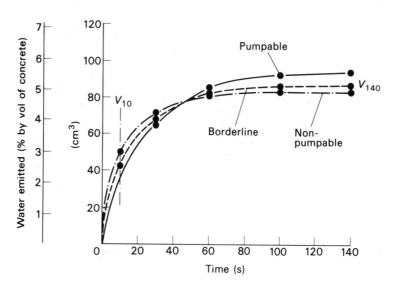

Figure 12.2 Typical bleed test results. (*Browne and Bamforth*)

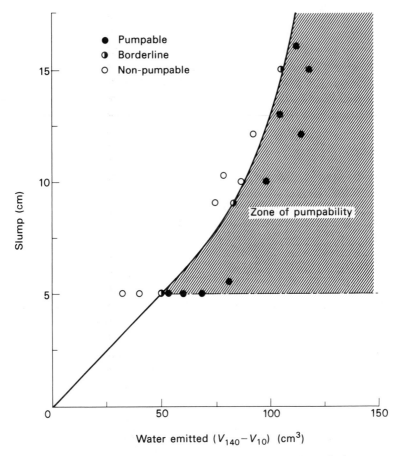

Figure 12.3 Interpretation of bleed test results. (*Browne and Bamforth*)

volume emitted after 10 s, V_{10}, could be taken as a measure of the proneness to dewatering. Minimum permissible values of $V_{140} - V_{10}$ were quantified for concretes with 20 mm aggregates over a range of workabilities, as assessed by slump, and the results were presented as in Figure 12.3.

The behaviour of concrete in relation to dewatering must of course be related to the void content. It is necessary to consider the void content of the combined coarse and fine aggregates, reduce it where possible by suitable choice of grading, and ensure that there is sufficient cement and finer fines to fill the voids. This is illustrated in Figure 12.4 which is due to Kempster[17]. Note that admixtures may be used as pumping aids (see Table 10.1 in Chapter 10) and, as pointed out in Chapter 11, a small addition of microsilica can be very effective in promoting pumpability.

There is, as yet, no published information on the relation between mix workability measured in terms of g (yield value) and h (plastic viscosity) and the pumpability as assessed either subjectively or objectively in terms of pumping pressures and flow rates. Work in this area is needed.

12.3 VIBRATION

It is well known that the effect of applying to fresh concrete a vibratory force, of nominal frequency around 50 to 300 Hz, is to increase its workability, and for decades it has been common practice to exploit this phenomenon on site to assist in the placing and compaction processes (Figure 12.1). Under the influence of vibration of a suitable amplitude and frequency, fresh concrete at first subsides and then appears to fuse or liquefy, so that it will flow relatively easily under its own weight, then follows a third stage during which air is seen to be expelled and the concrete compacts. The practical benefit is that either a given concrete may be handled more easily than it could otherwise be or, alternatively, the mix specification can be modified to achieve economies without loss of desirable properties of the hardened concrete. By using vibration it is possible to work with mixes drier than those suitable for hand placing and compaction and thus use a lower water/cement ratio to achieve higher final strength, or to use an unaltered water/cement ratio with a lower cement content and save on costs.

There are several different forms of equipment for application of vibration but perhaps the most common on site is the internal, or poker, vibrator which is actually immersed in the concrete. Use may also be made of vibrators attached to the shutters or formwork, or to a screeding bar. In the case of precast products, normal practice is to place the mould containing the concrete on a vibrating table, which may or may not be mounted on springs.

The mathematics of practical vibrating systems can be quite complicated but it is useful to consider the simplest form, which can be represented graphically by a sine wave. In this case, the vibrating mass moves from a mean rest position to an extreme position on one side and then back through the mean position to a symmetrically distant extreme position on the other side, and it continues to repeat this oscillating movement. An example is the movement of the bob of a simple pendulum. The distance of either extreme from the mean position is called the amplitude, A, and the number of complete cycles in one second is called the frequency, F or f, measured in cycles per second or Hertz (Hz). During the movement the mass has a changing velocity which is zero at each of the two extreme positions and reaches a maximum at the mean position but, conversely, the acceleration is zero at the mean position and maximum at the two extreme positions.

The movement can be represented graphically by a sine wave and is described by the equation

$$y = A \sin \omega t \qquad (12.1)$$

where y is the distance of the centre of mass from its mean position at time t, and ω is a constant. A simple vibration of this type is thus completely characterized by the values of the amplitude, A, and frequency, F, $(= \omega/2\pi)$ but it is often useful to consider also the derived quantities of maximum velocity, v, and maximum acceleration, a; it can easily be shown that they are given by

$$v = 2\pi AF \qquad (12.2)$$

$$a = 4\pi^2 AF^2 \qquad (12.3)$$

These are very simple relationships. The maximum velocity is proportional to the product of amplitude and frequency, while the maximum acceleration is proportional to the product of amplitude and the square of the frequency. Maximum acceleration is often expressed as a multiple of g_r, the acceleration due to gravity, and clearly, this quantity is obtained, in SI units, by dividing a by 9.81.

All the vibrators in common use in the concrete industry rely on the production of vibration by the rapid rotation of an out-of-balance weight and may therefore be referred to as rotating eccentric vibrators. Such an arrangement has the advantage of being simple and cheap, and is quite effective, but the waveform produced by real pieces of equipment is very complicated. The vibration is not in one direction only, the amplitude is variable, and many different frequencies are present. Any values of amplitude and frequency quoted for this type of vibrator are only nominal, and the latter is usually only the speed in rev/s of the eccentric mass. This does not matter for practical site use but it does if an attempt is to be made to elucidate what are the important factors for assessing the efficacy of vibration. Although in principle it is possible to express even a complicated waveform as the sum of a series of simple ones, it is highly desirable to use for investigational purposes a waveform that is itself simple, and this can be achieved by means of an electromagnetic vibrator. Such an apparatus was used in work to be described below.

The importance of the use of vibration for the placing and compacting of fresh concrete is reflected in the large amount of research that has been carried out, and in the fairly regular appearance of review and summary statements[18-21]. Experimental investigations may be roughly divided into two categories, those involving measurements on the fresh concrete, and those in which measurements, particularly of strength, were made on hardened concrete that had been cast under a variety of vibration conditions. Although it is of course the quality of the hardened concrete that is of ultimate importance, this latter approach is rather indirect, and it seems likely that a better understanding of the vibration process will emerge from direct studies on the behaviour of the fresh concrete.

Just as ordinary observation on site indicates that unvibrated fresh concrete possesses a yield value, because it can stand in a stable pile, so it also indicates that application of vibration reduces that yield value markedly, and at least to an extent such that the material can flow under the influence of its own self-weight. Experiments such as those of L'Hermite[22] 40 years ago support this conclusion, although his theory of the phenomenon has been shown to be invalid[23]. It is of course desirable to go further than this and a promising possibility is to attempt to obtain the flow curve of concrete while it is being vibrated. In fact, it is a simple matter to mount the bowl of the two-point workability apparatus on a suitable vibrating table so not only can the flow curve be obtained while the concrete is being vibrated, but also while it is not, and then the two flow curves can be compared, and this can be done for a whole range of applied frequencies and amplitudes.

Preliminary experiments of this type were carried out on a mix with an aggregate:cement ratio of 6:1 with 40% fines, in which all the cement had been replaced by pfa to reduce the rate of change with time, and water content had been adjusted to give 150 mm slump. The bowl of the workability apparatus was mounted on a table vibrator driven by a rotating eccentric formwork vibrator whose amplitude could be varied by altering the relative angular position of the eccentric masses. The concrete was placed in the bowl and the flow curve was obtained in the normal way when the vibrator was not operating. The impeller was then run at the top speed used and the vibrator was then switched on, resulting in a very rapid drop in torque followed by a slower rise. The rise was attributed to the occurrence of compaction, so the minimum torque observed was taken as a measure of the effect of vibration on workability and gave one point on the flow curve of the concrete under vibration. The concrete was then remixed, replaced in the bowl, and measurements were repeated at a lower impeller

	17	20	Void content – % by volume 25	28
15	PUMPED	PUMP BLOCKED DURING EXPERIMENT *	DID NOT PUMP *	DID NOT PUMP
20	PUMPED	PUMPED	DID NOT PUMP *	DID NOT PUMP
25	PUMPED	PUMPED	PUMPED	DID NOT PUMP *
30	PUMPED BUT DIFFICULT	PUMPED	PUMPED	DID NOT PUMP *

Cement content – % by volume

* mixes which became pumpable when cellulose ethers were added

Figure 12.4 Effect of voids in combined aggregate related to cement content as an indication of pumpability (*Kempster*). Sand content 35% of combined aggregate in all mixes. Water content adjusted to give 75 mm slump.

speed, to give another point on the flow curve. In this way four points were obtained and the flow curve was constructed.

The results obtained were of the form shown in Figure 12.5. The curve for the unvibrated concrete is the usual straight line but that for the vibrated concrete is quite different in that it is not linear and appears to start at or near the origin. Although experiments were carried out over a range of vibrator speeds (i.e. nominal frequencies) and amplitudes, no useful deductions could be made about the effects of frequency and amplitude because the waveform of the crude vibrator was so complicated. However, it was evident that the method was well worth pursuing if a more controllable vibrator could be obtained.

Suitable apparatus was immediately available only for smaller-scale work. A 100 mm bowl was mounted on a small electromagnetic vibrator and an impeller similar to that used in the workability apparatus, but scaled down in the same ratio as the bowl (2/5), was driven by a conventional viscometer mechanism which also provided the means of measuring the torque. Apparatus on this scale is not of course suitable for concrete or any material that contains large particles. Dimond[24] experimented on a variety of materials that had a range of rheological behaviours and obtained two important results:

(a) The effect of vibration was to produce a marked lowering of yield value and for materials that had no yield value vibration had no effect on the flow curve.

(b) When vibration did have an effect, that effect was immediate, and was also instantaneously reversible; in other words, switching on the vibrator caused an immediate drop in torque and switching it off caused the torque to return immediately to its original value.

Following Dimond's work, an investigation[25] was carried out with the same apparatus on cement pastes of water/cement ratios 0.28, 0.30 and 0.32, at frequencies from 25 to 100 Hz and at four different values of acceleration. It was found that for any impeller speed N, the torque under vibration, T_v, was related to the torque for the unvibrated material, T, by the simple relationship

$$T_v/T = K \cdot N^{kv} \tag{12.4}$$

where v is the maximum velocity of the vibration and K and k are constants for the particular experimental set-up. This means that the effect of vibration depended only on the maximum velocity of the vibration and not on the frequency and amplitude separately. It was also independent of the water/cement ratio of the paste and, as stated previously, was wholly and instantaneously reversible.

When an electromagnetic vibrator of sufficient capacity became available, it was possible to carry out similar experiments on concrete[26]. The apparatus used was essentially the same as in the preliminary experiments described earlier, except that the crude mechanical vibrator was replaced by the electromagnetic vibrator whose waveform was close to a sine wave and could therefore be characterized by a frequency and an amplitude, which could be controlled independently of each other. The set-up is shown in Figure 12.6. Two mixes were used; the cement was ordinary Portland, the aggregates were a 20 mm irregular gravel and a zone 2 sand, both air dried, the aggregate:cement ratios were $4\frac{1}{2}$:1 and 6:1 respectively, both with 40% fines. For each mix the water content to give a 75 mm slump was found by trial and error and then that water content was used in all subsequent batches of the mix, so the water/cement ratios were 0.47 and 0.57 respectively. The method for determining the flow curves was similar to that used in the preliminary experiments, except that for the vibrated concrete the number of points was increased from 4 to 7, and each test was completed in about 20 min. Experiments were carried out at six frequencies ranging from 15 to 100 Hz and at amplitudes corresponding to accelerations of $2\frac{1}{2}$, 5, $7\frac{1}{2}$ and 10 g.

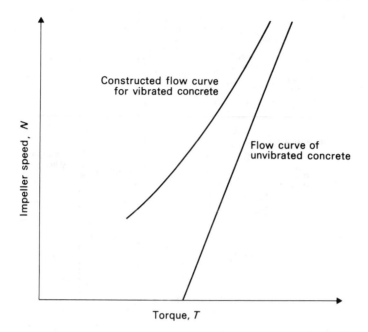

Figure 12.5 Effect of vibration on concrete flow curve.

In each case the effect of vibration was to change the flow curve from the simple straight line of the unvibrated concrete to a curve that seemed to start from the origin and could be represented by the equation

$$T_v = A_v N^{B_v} \tag{12.5}$$

where the suffix v indicates under vibration, and A and B are constants. This means that under vibration concrete behaves as what is known as a power-law pseudoplastic, and has zero yield-value.

Further analysis of the results led to an equation containing four empirical constants, by which the torque under vibration could be calculated from the values of g and h of the unvibrated concrete and the maximum acceleration and velocity of the vibration applied. The measure of agreement between the calculated and experimental values of T_v is shown for the first mix in Figure 12.7; 80% of the 276 points lie within $\pm 25\%$ of the line of equality. The agreement, while probably sufficient to justify the nature of the treatment given to the results, is not particularly satisfactory and, in any case, the equation concerned is too complicated for immediate practical use. Fortunately, interesting

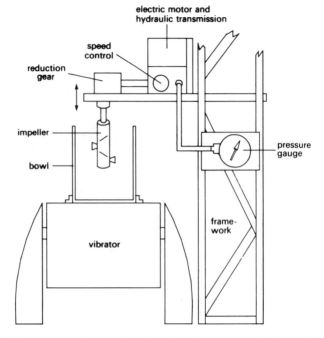

Figure 12.6 Experimental set-up for obtaining flow curve of vibrated concrete.

information can be obtained by considering only the early part of the vibrated flow curve.

In most normal practice in which vibration is used, the vibrated concrete is allowed to flow and compact under the influence of its own self-weight, and it does so at very low rates of shear. In the low-shear-rate region, that is at the lower end of the flow curve, the power-law pseudoplastic relationship can be approximated closely by a straight line passing through the origin. In other words, at these low shear rates the vibrated fresh concrete behaves as a simple Newtonian liquid whose fluidity ϕ (the reciprocal of the viscosity η) is simply proportional to the slope of the line, S, as shown in Figure 12.8.

Values of the slope, S, were obtained from the flow curves and it was found that at a given frequency S was simply related to the amplitude A by the equation

$$S = S_0(1 - e^{bA}) \tag{12.6}$$

where b is a constant at constant frequency. It was further found that the dependence of b on frequency f was of the very simple linear form

$$b = k(f - f_0) \tag{12.7}$$

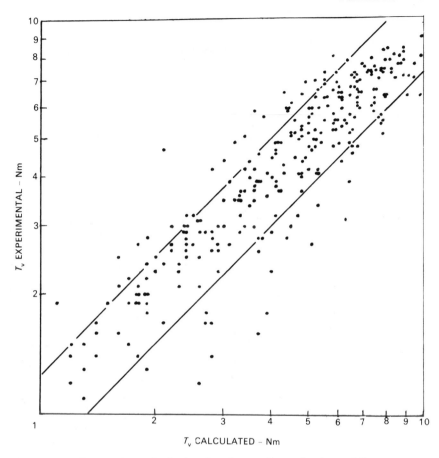

Figure 12.7 Comparison of calculated and experimental values of T_v.

where k is an empirical constant and f_0 is a small threshold frequency below which vibration has no effect on the concrete. Combining these two equations gives

$$S = S_0(1 - \exp(- Ak(f - f_0)))$$ (12.8)

Figure 12.9 shows that there is quite good agreement between the value of S calculated from this expression and the value of S obtained experimentally. The values of the threshold frequency f_0 for the two mixes were 12 and 16 Hz respectively. If the actual applied frequency is well in excess of f_0, equation 12.8 reduces to

$$S = S_0(1 - e^{-kAf})$$ (12.9)

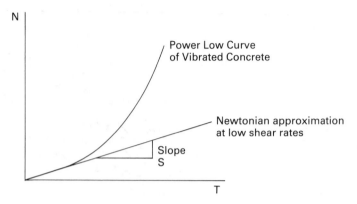

Figure 12.8 Linear approximation of power-law curve at low shear rate.

and, since Af is proportional to the maximum velocity, v, of the vibration this becomes

$$S = S_0(1 - e^{-k'v}) \qquad (12.10)$$

In words, the effectiveness of the vibration in promoting flow of the concrete depends only on the maximum velocity of that vibration.

Experiments of the type described are time-consuming and are not easy to carry out, but the finding that at low shear rates concrete under vibration behaves as a Newtonian liquid presents the possibility of designing a much simpler method[27]. The new apparatus is shown in Figure 12.10. A vertical pvc pipe, 700 mm long × 100 mm internal diameter, was mounted centrally inside a 356 mm diameter steel cylindrical bowl that was firmly clamped to the vibrating table. The lower end of the pipe was 100 mm above the base of the bowl but could effectively be lengthened to touch it by means of a sleeve sliding on the outside of the pipe. The apparatus was used as follows. The sleeve was lowered so that the extended pipe was in contact with the base of the bowl and the pipe was then filled with the concrete under test to a height of 600 to 700 mm. A light piston, connected to an ordinary builder's tape-measure by a string running over pulleys, was placed on the top of the concrete. The sleeve was then raised so that the lowest 100 mm of the concrete was unsupported (but because of its low workability remained in position), the vibrator was switched on at the chosen amplitude and frequency, and the downward movement of the concrete surface was monitored by reading the measuring tape against a fixed index.

Five mixes were investigated, and experiments were carried out at eleven different frequencies ranging from 16 to 200 Hz, and eight dif-

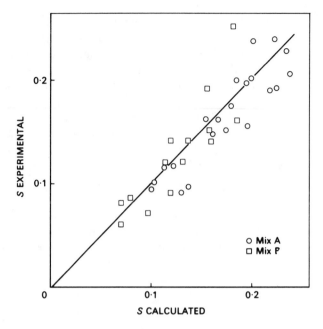

Figure 12.9 S experimental v. S calculated.

ferent accelerations ranging from 0.85 to 8.9g_r, that is, 88 different vibration conditions for each of the five-mixes. Air-dried aggregates were used and for the first batch of each mix the water content was adjusted to give a slump of 20–25 mm; that water content was then used in all subsequent batches of the particular mix.

The reasoning behind the design of this form of apparatus is that if fresh concrete under vibration behaves as a Newtonian liquid, its rate of flow out of the pipe will be simply proportional to the hydrostatic head. The rate of flow is proportional to the rate at which the height of the concrete column drops and the hydrostatic head is proportional to the height of the column provided the reasonable assumption is made that the drag at the pipe/concrete interface is either negligible or is proportional to the area of contact. Therefore

$$dH/dt = -b \cdot H \tag{12.11}$$

where the minus sign indicates that height H decreases as time t increases.

Integrating this equation gives

$$\ln H = \ln H_0 - bt \tag{12.12}$$

where H_o is the initial height of the concrete column. It follows that a plot of $\ln H$ as a function of time, t, should be a straight line with a slope of $-b$. It was found that all the experimental points fitted this relationship well over the first 15 s, so the efficacy of vibration may be studied by examining the relationships between b, which as already stated is a measure of the fluidity ϕ, and the vibration parameters.

Such a study showed, first of all, that there was a small threshold amplitude A_o below which, and an upper limiting frequency above which, vibration has no effect on the fresh concrete. The final expression obtained was of the form

$$b = (1/c) \cdot \ln(1 - f/F) \cdot (A - K_1(f/F)^{-K_2} + K_3) \qquad (12.13)$$

where F is the upper limiting frequency and c is an empirical constant that depends on the particular mix, while K_1, K_2 and K_3 are empirical constants that are the same for all the mixes. The good agreement

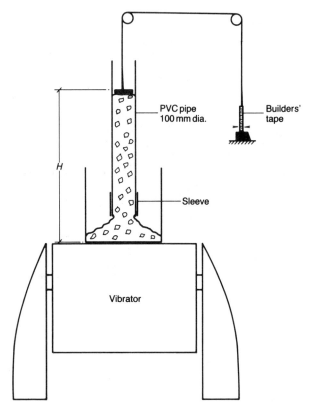

Figure 12.10 Experimental set-up for 'vertical pipe' experiments.

$$S = \tfrac{1}{4}[1 - \exp(-0.03Af)] \qquad (12)$$

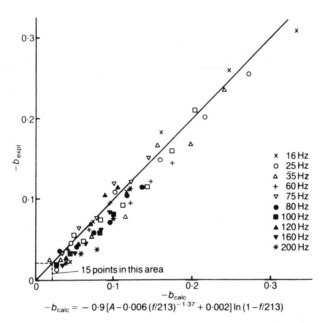

$$-b_{calc} = -0.9\,[A - 0.006\,(f/213)^{-1.37} + 0.002]\ln(1 - f/213)$$

Figure 12.11 Relationship between calculated and experimental values of b: mix A.

between the values of b calculated from this expression and the experimental values is shown in Figure 12.10. Equation 12.13 looks rather complicated but for values of amplitude A that are appreciably larger than the threshold amplitude A_o, and values of frequency that are appreciably less than the upper limiting frequency F, that is, for the practical range of conditions, the equation reduces to

$$b = \text{constant} \times Af = \text{constant} \times \text{velocity } v \qquad (12.14)$$

that is, once again, the effect of vibration depends only on the maximum velocity.

In fact, over the whole range the value of b is almost constant at any constant value of velocity v. Figure 12.11 shows the calculated value of b plotted against frequency for various constant values of velocity and various constant values of acceleration. It can be seen that b depends markedly on frequency at a fixed acceleration but is, as just stated, almost independent of frequency at constant velocity.

Limitations to the conditions of vibration that can be applied to the concrete are of course imposed by the design of the apparatus used. For the vibrating table used in this work the amplitude cannot exceed 12 mm, even at the lowest frequencies, and the acceleration cannot exceed $10\,g_r$. The corresponding limiting lines are also shown in Figure 12.12 and it is evident that, using this particular apparatus, the best results, that is, the highest values of b, are obtained by working in the region of the cusp formed by the two limiting lines; in this case that means at a frequency of something under 20 Hz, which is considerably lower than the nominal frequency of most commercial vibrating tables.

A further test of the justification for regarding the behaviour of vibrated concrete at low shear rates as Newtonian, and the validity of the arguments used in establishing the various equations, can be made by comparing the results from the two quite different types of experiment. If the reasoning has been satisfactory the two quantities, S from equation 12.8, and b from the pipe experiments, should be simply related, for the same, mix, because each of them is a measure of the fluidity ϕ.

Because in the earlier work it was necessary to use concretes of slightly higher workability, there is in fact no case of exactly the same

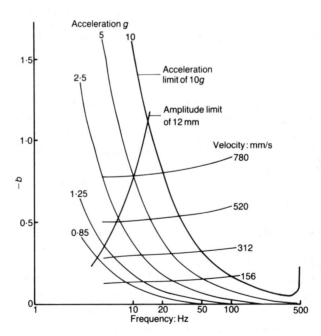

Figure 12.12 Fluidity expressed as b (calculated) as a function of frequency f at constant values of acceleration and velocity.

mix being used in both sets of experiments, but two mixes which differed only in water content were used. For these two, it is still reasonable to expect a relationship between S and b although some departure from linearity is likely. Figure 12.13 shows that there is a very good correlation and this result of a very good agreement between measures of fluidity obtained from two very different types of experiment is very encouraging.

The important practical conclusions from this work may therefore be summarized as follows:

(a)   At low shear rates, which are the ones important in practice, fresh concrete under the influence of vibration behaves as a simple Newtonian liquid.

(b)   There is a threshold amplitude below which, and an upper limiting frequency above which, vibration has no practical effect in that the yield value is not reduced to a level low enough for flow to occur.

(c)   The important parameter for assessing the effectiveness of vibration is the maximum velocity of the vibration.

It may also be noted that the results from the first set of experiments, on the determination of the full flow curve, show that at high rates of shear the vibrated curve crosses the unvibrated curve so above this

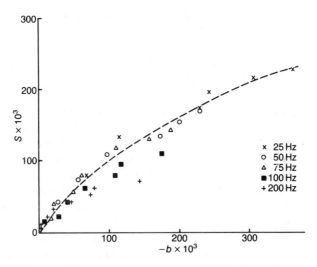

Figure 12.13 Relationship between measures of fluidity from flow curves S and pipe experiments b.

crossing point vibration actually **reduces** workability. At present, insufficient is known about the effective average shear rates prevailing in practical processes, but it is quite possible that the use of vibration in conjunction with some other mechanical process would actually be disadvantageous. The results also suggest that some reconsideration of the design of vibrating tables for maximization of vibration velocity might be worthwhile.

Although the conclusions are likely to have some general validity, they should for the time being be regarded as applying specifically to table vibrators. In the case of internal vibrators, for example, other factors are likely to be important, because the effectiveness of the vibration decreases with distance from the source, that is, it is attenuated. The practical manifestation of this attenuation is that an internal vibrator, or poker, has a radius of action which can be seen by ordinary visual observation and, as a consequence, a poker must be inserted at about 400 mm centres. Withdrawal of a poker must of course be performed while the poker is still vibrating, otherwise the yield value of the concrete will immediately be restored and withdrawal will leave a hole.

It is possible that work of the kind described will eventually lead to proposals for an improved workability test for concretes of extremely low workability but no suggestions can be made at present. The fluidity, and therefore the workability, of a vibrated concrete depends not only on the properties and proportions of the mix constituents but also on the parameters of the applied vibration. Further experiments are needed, on concretes of a wide range of workabilities, to investigate the relationships between the workability parameters, g and h, of the unvibrated material and, for example, the constant in equation 12.14.

A limited amount of other work has already been stimulated. Kakuta and Kojima[28–29] used a set-up essentially the same as that shown in Figure 12.6 except that the uniaxial form of the two-point apparatus was replaced by the planetary form and it was mounted on a vibrating table that was actuated by a rotating eccentric. The nominal frequency (i.e. speed of rotation of the eccentric) was 1750 rev/min (nominal 29 Hz) and was not varied, but the amplitude was set at four different levels. No information is given about the nature of the waveform but accelerometers indicated that there was an appreciable horizontal component and that the vertical component ranged from about $1 g_r$ to about $5 g_r$, (as measured by the accelerometers).

The main results of this work were to confirm the early findings of the work described previously: that is, that the linear curve of unvibrated concrete was transformed by vibration to a power-law curve, although Kakuta and Kojima do suggest that over the full range of shear rates a better fit is obtained by using a power law for the lower half of the curve and a straight line for the upper portion. The correlation coefficients they quote do not seem to give much support to this suggestion.

There is a fruitful field for investigation using either the original form of the apparatus, or the very simple vertical-pipe apparatus which is so easy to use that large numbers of variables can be coped with. In the meantime, it might be thought that some justification has been provided for the use of the Vebe consistometer but such a conclusion would be quite wrong. As already stated, the waveform associated with that apparatus is so complicated and variable as not to be readily characterized, and there can be no guarantee that results from such an arrangement relate in any simple or useful way to the behaviour of concrete vibrated under site or factory conditions.

There is not as yet any satisfactory theory of the effect of vibration on fresh concrete, and there does not seem to have been any serious attempt to provide one since that of L'Hermite[22] already mentioned and criticised[23]. Recently, Chandler[30] has said that the mechanism can be explained easily in terms of pore pressure, using a soil mechanics approach, and suggests that the implication is that observed frequency effects may only be a result of resonance in a particular mould. Clearly, this is another area where further work is needed.

12.4 REFERENCES

1. Quality Scheme for Ready Mixed Concrete, *Manual of Quality Systems for Concrete*. The Quality Scheme for ready mixed concrete, Walton-on-Thames, May 1984, 23pp and later edition 1989.
2. Stubbs, H.A. (1977) The ramifications of truck mixed concrete, Project Report on Advanced Concrete Technology Course of the Cement & Concrete Association.
3. Sym, R. Private communication to G.H. Tattersall, 23 October 1989.
4. Beitzel, H. (1980) The influence of mixing time on concrete quality, *World Construction*, Sept.
5. Beitzel, H. (1981) The influence of mixing time on the quality of concrete, *Hong Kong Contractor*, January.
6. Beitzel, H. (1982) Random sample analysis for the assessment of concrete mixers, *Betonwerk u. Fertigteil-Technik*, 606–14.
7. Anon. (1985) Productivity gain in mixer design and maintenance *World Construction*, January, 20–1 and 4–5.
8. Beitzel, H. (1983) Einsatz von Betonmischern, *Schweizer Ingenieur und Architekt*, 965–73.
9. Beitzel, H. (1988) Concrete production plants and mixers some aspects of their design and operation. Betonwerk – Fertigteil – Technik. 234–8, 305–10, 822–4 and 1985, 667–72.
10. Morinaga, S. (1973) Pumpability of concrete and pumping pressure in

pipelines, in *Fresh Concrete: Important Properties and their Measurement, Proceedings of RILEM Seminar held 22–24 March 1973, Leeds*, Vol. 3, Leeds, The University, pp. 7.3-1–7.3-39.

11. Sakuta, M. *et al.* (1979) Pumpability and rheological properties of fresh concrete, in *Proceedings of Conference on Quality Control of Concrete Structures, 17–21 June 1979, Stockholm*, Vol. 2, Stockholm, Swedish Cement & Concrete Research Institute, 125–32.

12. Aleekseev, S.N. (1952) On the calculation of resistance in the pipes of concrete pumps, *Mekhanizatsiya Stroitel'stva*, **9**(1), 8–13, (Translated by K. Simon, Building Research Station Library Communication No. 450, 1953).

13. Weber, R. (1963) The transport of concrete by pipeline (Translated by C. van Amerongen, Cement & Concrete Association Translation No. 129, 1968.)

14. Ede, A.N. (1957) The resistance of concrete pumped through pipelines, *Magazine of Concrete Research*, **9**(27), 129–40.

15. Tattersall, G.H. and Banfill, P.F.G. (1983) *The Rheology of Fresh Concrete*, London, Pitman, Chapter 4.

16. Browne, R.D. and Bamforth, P.B. (1977) Tests to establish concrete pumpability, *Journal of the American Concrete Institute, Proceedings*. **74**, 193–207.

17. Kempster, E. (1969) Pumpable concrete, *Contract Journal* **229**(4693), 605–7; **229**(4694), 740–1; Building Research Station Current Paper 29/69.

18. Institution of Civil Engineers and Institution of Structural Engineers, (1956) *The vibration of concrete*. Report of a Joint Committee, 64pp.

19. Murphy, W.E. (1964) A survey of post-war British research on the vibration of concrete, London, Cement & Concrete Association, Technical Report TRA 382, 25pp.

20. ACI Committee 309 (1981) Behaviour of fresh concrete during vibration, *Journal of the American Concrete Institute, Proceedings*, **78**(1), 36–53.

21. American Concrete Institute (1987) in Gebler, S.H. (Ed.) *Consolidation of Concrete*. Detroit, 1987. Special Publication SP-96, p. 250.

22. L'Hermite, R. (1948) The rheology of fresh concrete and vibration, Publication
Technique No. 2, Paris, CERILH (Library Translation Cj. 9, London, Cement and Concrete Association, 1949.)

23. Tattersall, G.H. (1954) The rheology of cement pastes, fresh mortars and concretes,
M Sc Thesis, University of London.

24. Dimond, C.R. (1980) Unpublished internal report, Department of Building Science
University of Sheffield. Reported in Reference (9), 301–2.

25. Tattersall, G.H. and Baker, P.H. (1988) The effect of vibration on the rheological
properties of fresh concrete, *Magazine of Concrete Research*, **40**(143), 79–89.

26. Tattersall, G.H. and Baker, P.H. (1989) An investigation on the effect of vibration on the workability of fresh concrete using a vertical pipe apparatus, *Magazine of Concrete Research*, **41**(146), 3–9.

27. Kakuta, S. and Kojima, T. (1989) Effect of chemical admixtures on the rheology of fresh concrete during vibration in *Proceedings of 3rd International Conference on Superplasticizers and other chemical admixtures in concrete, 4–6 Oct. 1989, Ottawa*. Session 3.

28. Kakuta, S and Kojima, T. (1990) Rheology of fresh concrete under vibration

in Banfill, P.F.G. (Ed.), *Proceedings of Conference of Beitish Society of Rheology on Rheology of Fresh Cement and Concrete, held Liverpool 16–29 Mar. 1990*, E. & F.N. Spon, London, 339–42.
29. Chandler, H.W. (1990) Pore fluid pressure in ceramic processing: a soil mechanics approach, *British Ceramic Transactions*, **89**(5), 153–8.

13 Specification of workability

For any particular job it is necessary to know what workability is required so that a specification may be written, so, obviously, unless the decision is to be based on personal practical experience, information relating workability levels to various particular types of job is needed. *Road Note 4*[1], which for many years was the best-known UK guide to concrete-mix design, gave such information in tabular form and listed four categories of workability in purely descriptive terms ranging from 'very low' to 'high'. With each category there was associated a corresponding compacting factor, a range of slump values, and examples of jobs for which that grade of workability was said to be suitable. A caveat was entered in a footnote which said: 'The slump is not definitely related to the workability or the compacting factor. The figures given must, therefore, be regarded as a rough indication of the order of the slump and nothing more'. The value of attempting to give general rules at all has more recently been questioned, and in *Design of Normal Concrete Mixes*[2], which was produced as an intended replacement for *Road Note 4*, it is stated, as was pointed out earlier, that 'it is not considered practical for this Note to define the workability required for various types of construction or placing conditions since this is affected by many factors'.*

In practice therefore, workability specifications are devised on the basis of experience, and are stated in terms of results from one of the standard tests. In this connection, by far the most commonly quoted test is the slump test; the flow table is occasionally quoted, compacting factor rarely, and the Vebe hardly ever. Of course, as with any specification, it is necessary to quote tolerances on the desired value and this is done in British Standard BS 5328:1981 *Methods for specifying concrete*, where it is stated that workability shall be within the limits given in Table 13.1.

---

*See Appendix.

Table 13.1 Tolerances on workability specification as given in BS 5328 : 1981*

Specified test	Specified value	Tolerance
Slump	All	±25 mm or $\frac{1}{3}$ of specified value, whichever is the greater
Slump with sampling from early part of discharge	25 mm 50 mm 75 mm+	$+35$ mm, -25 mm ±35 mm $\pm(\frac{1}{3}$ of specified $+$ 10 mm)
Compacting Factor	0.90+ 0.81 to 0.89 0.80$-$	±0.03 ±0.04 ±0.05
Vebe	All	±3 s or $(\frac{1}{5}$ of specified) whichever is the greater

*See Appendix.

These requirements are not always applied in practice. Figure 13.1 shows slump-test results obtained on 39 nominally identical batches delivered to a typical (but not necessarily representative) site during one day. The specified slump was 50 mm and, since samples were taken from the early part of the discharge, the tolerance according to BS 5328 should have been ±35 mm so the seven batches whose slumps exceeded 85 mm should have been rejected. In fact none of them was, but two batches whose slumps were 25 mm and 35 mm were rejected when, according to the specification, rejection was not justified.

The limits given in Table 13.1 are very wide, particularly in the case of slump, which as remarked earlier, is by far the most commonly quoted. For example, if the sample is taken from the early part of the discharge, as is common on site and may be the only procedure practically possible, a specified 25 mm slump mix must be accepted if the test result lies between zero and 60 mm, a 50 mm if it lies between 15 and 85 mm, and 100 mm if it lies between 57 and 143 mm. These ranges imply the possible acceptance not only of a wide range of workabilities, including what may be unsuitable, but also of concretes whose composition may be incorrect in a way that will show up in cube tests a month later.

It could be argued that there is an inconsistency in quoting workability tolerances as definite ranges when cube-test results are considered on a probability basis, because both workability and strength results are subject to the same sort of variability arising from manufacturing and testing variability. However, whichever way it is done,

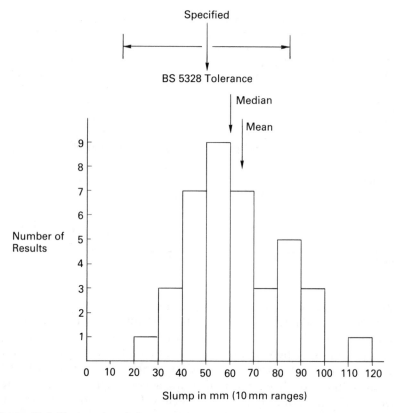

Figure 13.1 Site results of slump testing on 39 nominally identical batches of concrete (specified slump = 50 mm).

the establishment of tolerances properly requires consideration of both manufacturing variability and testing variability.

Sym[3] has approached this problem by accepting the tolerances as given by BS 5328 and then considering whether the variability associated with a particular test is low enough for that test to be suitable for use as intended in BS 5328. He proposes a measure which he calls the **Test Capability Index** defined as

$$I = 2d/s \qquad (13.1)$$

where the material is required to meet a specification of $T \pm d$ and S is testing standard deviation. He considers that a value of $I = 4$ or less indicates that the test is unsatisfactory for the tolerances proposed while a value of $I = 8$ or more indicates it is satisfactory. Because the

*See Appendix for flow table.

values of I for slump and compacting factor lie between these he concludes that the case for both those tests is a marginal one.*

The meaning of I may be explained as follows. If $I = 4$, the whole of the variability allowed in the tolerance range is required for test variability and nothing is left to allow for process variability. In other words, if all the batches had an actual slump equal to the specified value, 1 in 20 of the test results would lie outside the tolerances simply because of testing variability. A value of $I = 8$ is equivalent to saying that the testing standard deviation shall be not more than half of the standard deviation associated with the whole operation of making the concrete and testing it, and with the same probability of a 1 in 20 failure rate. The choice of this condition is arbitrary. While the Test Capability Index may be a useful quantity to consider it will be realized that it is somewhat inaptly named because it is not, as may seem to be implied, a characteristic of the test alone; its value depends on the chosen tolerance too.

If the desirability of a value of I not less than 8 is accepted, and if calculation is based on the value of 11 mm for testing standard deviation, argued for in Chapter 2, the tolerance on slump values would have to be widened to around ±40 mm, to restrict failure to 1 in 20.

Although the histogram of site results in Figure 13.1 shows some evidence of skewness it is probably close enough to a normal distribution to permit calculations to be made on that basis without serious error, as suggested by the fact that the mean at 66 mm is close to the median at 60 mm. The standard deviation, calculated as that of a normal distribution, is 20 mm and the corresponding confidence limits at the 0.05 level are $L = ±40$ mm. That means that 1 in 20, or 2 of the 39 results, can be expected to lie outside the range $65 ± 40$ mm and inspection of Figure 13.1 shows that such is the case: one at the lower end and one at the upper end.

The agreement between these confidence limits and the suggested necessary extended tolerances (i.e. both at 40 mm) is coincidental. More work is needed to resolve this matter satisfactorily but, so far as total variability is concerned, it is likely that much information already exists and only needs collecting.

13.1 SPECIFICATION IN TERMS OF THE TWO-POINT TEST

Just as it has been accepted that specification of workability levels in terms of mean values obtained by standard tests can be done only on the basis of experience, so also must the determination of values of g (yield value) and h (plastic viscosity) appropriate for particular jobs

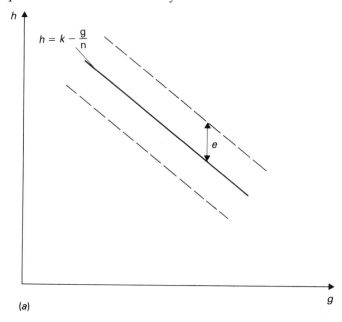

$$h = k - \frac{g}{n}$$

(a)

Figure 13.2 Suitability band of concretes for particular job.

be done from experience in practice, but it can be done in a systematic way. Moreover, when the desired values and ranges are known, the situation does not become clouded by the need to use overall values for test accuracy, as for the standard tests, because an experimental error can be quoted for every separate measurement, as already explained in Chapter 5.

First of all, it must be recognized that any attempt to associate suitability of concrete for a particular job with any form of quantitative measurement can only be carried out on actual jobs; it cannot in principle be done in the laboratory, and this applies to the two-point test as it does to any other. In other words, before any answer can be given to the question 'What values of g and h are needed for this job?' it is necessary to carry out site trials in which measurements are made on as many batches as possible and, on the same batches, appropriate assessments of the suitability of the concrete are recorded. Those assessments may sometimes be made in terms of objective measurements, such as, for example, pressures needed for pumping, but more often they will have to be made on a subjective basis.

It is possible to make some progress by recording simply whether the concrete was satisfactory or not, and in early site trials subjective assessments were made in simple descriptive terms such as 'a bit stiff'

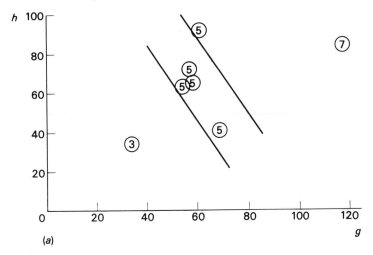

(a)

Figure 13.3 Site results showing concretes within suitability band. (Figures indicate suitability rating on subjective scale.)

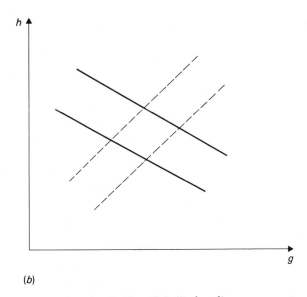

(b)

Figure 13.4 Suitability band as in Fig. 13.2. Broken lines represent approximate relationship between g and h when only water content varies. Acceptable concretes lie in the 'diamond' common to both bands.

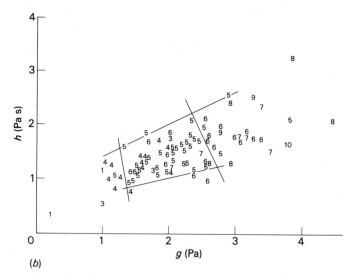

(b)

Figure 13.5 Site results on diagram similar to Figure 13.4. (Figures indicate suitability rating on subjective scale.)

and so on. It was soon realized that this was rather unsatisfactory, so in later trials the site engineers were asked to allocate a number to each batch of concrete on a scale of 1 to 10, where 5 meant perfect for the job, higher numbers meant increasing difficulty because workability was too low, and lower numbers meant that workability was higher than necessary.

Now, it is reasonable to suppose that associated with every practical job there is some (unknown) effective average shear rate and it follows that any two concretes whose apparent viscosities are the same at that shear rate will behave in a similar way on the job. Apparent viscosity is defined as shear stress divided by shear rate so a measure of it, k say, in terms of two-point test results, is given by torque divided by speed or

$$k = T/N = g/N + h \tag{13.1}$$

If $N = n$ is the speed in the apparatus that corresponds to the shear rate on the job it follows that any two concretes whose values of

$$k = g/n + h \tag{13.2}$$

are the same, will behave in the same way on the job, so that if one of them is satisfactory the other will be too, even though their separate values of g and h may differ. Equation 13.2 may be rewritten as

$$h = k - g/n \tag{13.3}$$

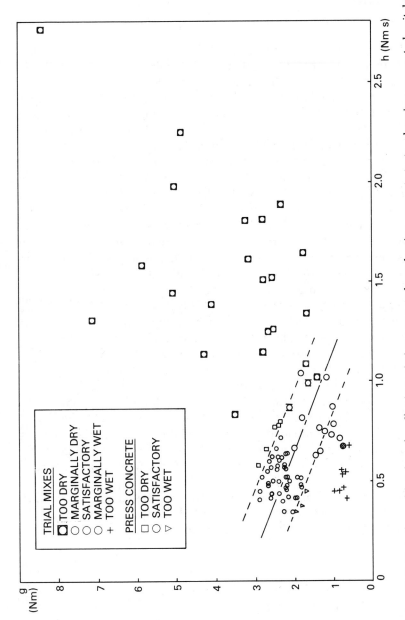

Figure 13.6 Plot of *g* and *h* values for all mix-variation tests and production concrete tests, showing suggested suitability band for the process (*Kay*).

which is the equation of a straight-line relationship between g and h with an intercept k on the h axis and a slope of $-1/n$. If a tolerance $\pm e$ is allowed on k, this equation may be modified to

$$h = k - g/n \pm e \tag{13.4}$$

which represents a band, as shown in Figure 13.1. Therefore, on a plot of h against g, the points for all concretes suitable for the job should tend to fall in a band sloping down from left to right.

Figure 13.3 shows the few results obtained on a site where a super-plasticized high-workability concrete was being used for casting a basement, and, while this amount of evidence is insufficient to prove the argument given, it is at least compatible with it.

An attempt was made to obtain many more points on another job but unfortunately, for this experiment, it turned out that the only factor causing variability of workability was variation in water content. In this case, there is a strong correlation between g and h so the satisfactory concretes will lie in the area of the g v. h relationship that is common with the 'suitability band' as shown in Figure 13.4. Figure 13.5 shows the results obtained on this site, where high-workability concrete was being used for piles, and again, it can be seen that they fit reasonably well to the scheme suggested.

The most comprehensive study of this type was carried out by Kay[4] on concretes to be considered for use in the hydraulic pressing of slabs. He investigated trial mixes at three levels of aggregate:cement ratio, three levels of percentage fines, and five levels of water/cement ratio, that is, 45 trial mixes, and he also obtained results on production mixes. In addition to measuring g and h for each mix he obtained two other assessments of suitability, one on a too-wet to too-dry scale, as judged by the experienced press operator, and the other on a scale of judgement of the finished slab in terms of edge and surface defects. As a result, he was able to produce the diagram shown in Figure 13.6 illustrating quite clearly the suitability band for concretes for the process he studied.

Further, Kay was also able to estimate that the value of n in equation 13.4, for the process he was considering, was 1.94 rev/s and the required apparent viscosity at that shear rate, expressed in terms of k, was

$$k = 1.7 \pm 0.3 \, \text{Nms} \tag{13.5}$$

These figures were actually obtained by a consideration of normal production mixes but have been applied[5] to the data from the 45 trial mixes to give the information shown in Figure 13.7. This figure is a pseudo-three-dimensional plot which shows the dependence of apparent viscosity at the process shear rate on each of water/cement ratio, fines content, and aggregate:cement ratio, while the other two are constant at various levels, and it also shows how those dependences change as the other two factors change. To avoid excessive complication of the drawing, only four of the aggregate:cement ratio lines have actually been drawn in. The range of acceptable apparent viscosities is also shown so it is easy to pick out the various mix compositions that would yield satisfactory results. A diagram of this type can also be used for quality-control purposes.

Exactly the same techniques can be used for any other site or works production process in which fresh concrete is used but, of course, such work will require effort from the industry itself because it cannot be done in the laboratory.

13.2 REFERENCES

1. Anon. (1950) *Design of Concrete Mixes, Road Note 4*, Road Research Laboratory, London, HMSO.
2. Teychenné, D.C. *et al.* (1988) *Design of Normal Concrete Mixes*, reviewed edition, Building Research Establishment, 43pp.
3. Sym, R. (1988) Precision of BS 1881 concrete tests. Part 2: Assessment. Slough, British Cement Association, 13pp.
4. Kay, D.A. (1987) The workability of hydraulically pressed concrete, Project Report, Advanced Concrete Technology Course, Cement & Concrete Association.
5. Tattersall, G.H. (1987) Workability measurement and its application to the control of concrete production, in Wierig, H. (ed.) *Proceedings of Fachkolloquium 'Zementleim, Frischmortel, Frischbeton'* at University of Hannover, 1987 Oct. 1–2 Hannover, the University, 88–98.

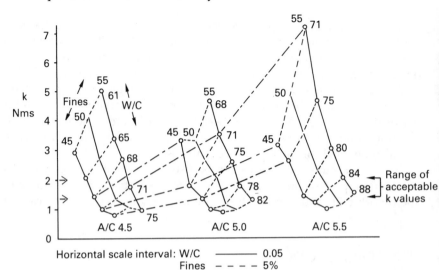

Figure 13.7 Dependence of apparent viscosity, at process shear rate, on water/cement ratio, fines content, and aggregate; cement ratio (*Diagram prepared from Kays' results*).

Explanation of Figure 13.7

Figure 13.7, which looks rather complicated, is plotted on a basis described by McIntosh*, and can most easily be understood by considering how it was built up.

Kay found that the average effective shear rate of the hydraulic pressing process was equivalent to a speed of 1.94 rev/s in the two-point apparatus. The value of k, apparent viscosity, at this shear rate has been calculated for each of his mixes by substituting in equation 13.1:

$$k = g/1.94 + h$$

and this calculated value has been plotted on the vertical axis.

All three of the independent variables, (W/C ratio, % fines, A/C ratio) have been plotted on the horizontal axis.

Consider first of all the results for an A/C of 4.5 and of these, those for 45% fines. The relevant five points for k as a function of W/C are plotted in the normal way with k on the vertical axis and W/C on the horizontal axis with a scale such that each 10 mm represents a change in W/C of 0.05. This gives the first line in the figure, the one at the extreme left.

The next line to be plotted is for k v. W/C for 50% fines. This is plotted exactly as the first line except that the origin is shifted horizontally by 10 mm to represent the 5% change in fines. This gives the second line from the left. Similarly the third line, for 55% fines, is plotted with another 10 mm shift to represent the further 5% change in fines.

Thus, the three full lines show how k changes with W/C at three fines contents but, in addition, because these lines are displaced relative to each other by distances proportional to the change in fines, points may be joined as by the dotted lines to show how k changes with fines content at constant W/C.

The other results, for A/C of 5.0 and 5.5, are plotted in a similar way but with the whole lot for a given A/C shifted by a distance proportional to the change in A/C. Points may then be joined as by the chain lines to show how k changes with A/C. Only three of these lines have been drawn otherwise the diagram would have become impossible to read. That difficulty can be overcome by using colours.

The advantage of this method of plotting is that it shows not only how k depends on each of the independent variables separately when the other two are constant, but also shows how those dependences change if the other two are not held constant. For example, the part of the diagram referring to A/C of 4.5 shows how the dependence of k on W/C depends itself on the value of fines, and how the dependence of k on fines depends on the value of W/C. It is also easy to interpolate; e.g. k for a mix with A/C 4.5, 52% fines and W/C 0.67 will be about 2.7 Nms.

Kay found that mixes suitable for hydraulic pressing must have $k = 1.7 \pm 0.3$ at the shear rate corresponding to $n = 1.94$. This 'acceptable range' is also shown in Figure 13.7. All mixes whose compositions lie between the parallel lines indicated by the arrows will perform well in the process. Any whose composition is outside those lines will perform less well, and more so the further they are away from the band enclosed by the parallel lines.

* McIntosh J.D. (1949) Method of graphing several variables *Magazine of Concrete Research*, **3**, 145–8

14 Workability measurement as a means of quality control

14.1 INTRODUCTION

Quality control is an important consideration in all but the most trivial manufacturing processes and its importance for concrete production has increased with the advent of quality assurance schemes and the introduction of the concept of what are known as 'designated mixes'. Tipler[1] has reviewed the efforts to improve concrete quality over a period of 30 years or so, up to the introduction of the Quality Scheme for Ready Mixed Concrete (QSRMC) in 1984, and the publication of its *Manual*[2]. Since, as Newman[3] points out, the ready-mixed concrete industry is the source of the bulk of site-placed concrete (two-thirds in 1986) this is a significant development, and the situation now is that QSRMC employs staff to assess continually over 1100 plants which together produce over 90% of UK ready-mixed concrete.

In 1989 there occurred improvements in that, as stated by Barber[4], 'The governing board became independent with a broad representation from industry, government and the private sector. In parallel with this, certificated companies were required to introduce full BS 5750 quality systems into their operations'.

The designated concrete-mix system has been described by Harrison[5] and the intention is that a purchaser need only select an appropriate mix title and state whether the intended use is for unreinforced, reinforced or prestressed concrete. For example, designated mix RC40 with a recommended slump value of 75 mm is listed as suitable for reinforced concrete to be subjected to severe exposure; the characteristic strength would be 40N/mm^2, it would have a minimum cement content of 325kg/m^3 and a maximum water/cement ratio of 0.55. The responsibility for meeting requirements would lie with the fresh concrete producer and Harrison envisages that 'Given time, reduced or no site testing will become the norm for most situations . . .', but he recognizes that purchasers need to develop confidence in the designated mix system and may wish to reduce the level of site testing progressively. Detailed mix specifications are to be included in BS 5328.

According to Barber[4]: 'QSRMC recommends that a single clause (in specifications) is sufficient and effective to provide a basic specification covering materials, acceptance procedures with control of strength and durability. That single clause is:
READY MIXED CONCRETE:
"Ready-mixed concrete shall be supplied from a plant currently certificated by the Quality Scheme for Ready Mixed Concrete, or a plant certificated under a Product Conformity Certification Scheme accredited to a standard equivalent to that of the National Accreditation Council for Certification Bodies"'.

14.2 GENERAL PRINCIPLES OF QUALITY CONTROL

Quality control of any manufactured product may be achieved through the following stages:
(a)   specification of raw materials;
(b)   testing of raw materials and use of the results to modify (c) below;
(c)   specification of processing;
(d)   testing at intermediate stages of production and use of the results to modify (c) above, to modify the product at the intermediate stage, or, if necessary, to reject the product at this stage and avoid further financial loss;
(e)   testing of the final product for pass/fail and use of results to modify (c) above.

The emphasis to be laid on each of these five steps will depend on the natures of the process and the product, and on the degree of control that is thought to be acceptable. In a very simple process it may be sufficient to carry out (a) and (c) only but in the case of concrete it is necessary to consider all five, and the extent to which this has been done in the *QSRMC Manual* has been discussed elsewhere[6,7,8].

14.3 PRESENT PRACTICE

The usual way in which account is taken of the five steps in concrete production is as follows.

Step (a)

Specifications for raw materials are laid down in terms of the appropriate standards but, as is well known, that cannot be sufficient to ensure adequate constancy. Complaints about variability of the water requirement of cement are common and the properties of nominally identical aggregates can vary significantly, particularly in their gradings, the proportion of crushed oversize material, and of course, moisture content. It would not be practically possible to tighten up the specifications to eliminate these variations so it is necessary to consider whether the alternative of corrective action can be applied, and this involves a consideration of the later steps.

Step (b)

Testing of raw materials may be carried out to ensure that they comply with the specification, and also to look at variations within the permitted limits with a view to modifying step (c). However, when this is done it is only on a relatively long-term basis so that only trends can be allowed for; there is no attempt to assess raw materials for each batch of concrete to permit appropriate adjustments to be made. It would be very difficult to institute such testing at an adequate level; for example, grading and particle shape of aggregates would not only be difficult to assess on a more or less continuous basis but there is no way of describing the results of a test in sufficiently simple terms.

The one important factor that, in principle, could be measured and allowed for is aggregate moisture content, but even here there are difficulties that arise not so much in devising a method of measurement as in ensuring that the result is representative, as has been pointed out by Anthony[9]. Even on a laboratory scale it is not easy. Although it is claimed for some modern computer-controlled plants that the difficulties have been overcome, the normal practice is to assume an average value for moisture content and use it in calculation of batch weights of aggregates and water. The frequency and efficiency of checking on the assumed value will vary from plant to plant but, clearly, here is a probable source of significant variability in the product.

Step (c)

Specification of processing involves laying down requirements for the design of batching and mixing plant and of tolerances for the accuracy of measuring equipment, but it should also include standardization of the mixing procedure, particularly if the mix contains an admixture, when even total time of mixing may be important.

It may seem a simple matter to control the quantities of the components of a mix. For example, cement can be weighed accurately

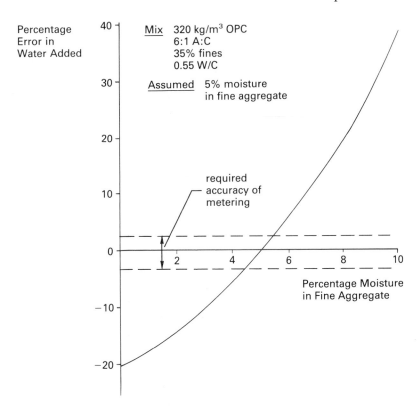

Figure 14.1 Possible errors in quantity of water added.

enough but, even here, problems can arise from equipment failure. A case has been reported[10] where the aggregate:cement ratio drifted from 6.5:1 to 8:1 because the batching equipment was weighing cement stuck in it as well as the cement discharged to the mixing drum, and there have been several instances of production of batches with low, or even zero, cement content because of a leakage in the wall between cement and cement-replacement silos.

Problems can also arise from human error even in the most sophisticated plant. Limestone aggregate may be used instead of a specified gravel and on one important job a very expensive mistake was made when a large quantity of concrete was made with the wrong admixture (an air-entraining agent instead of a plasticizer), and the error was discovered after the material had set, in heavily reinforced sections.

It may be thought that difficulties of this type are so infrequent as to be unimportant, but when they do occur they may be very costly.

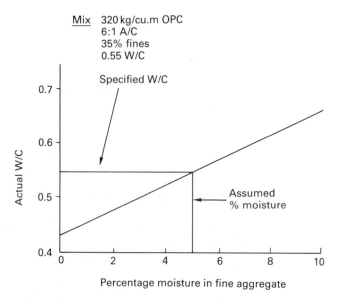

Figure 14.2 Possible errors in water/cement ratio.

It may also be said, with hindsight, that they were due to carelessness and should not have happened at all, but such a criticism is facile, although the importance of good supervision should be recognised.

A more pressing problem, because it is present all the time, is the one arising from the inadequate knowledge of aggregate moisture content mentioned earlier. Figures 14.1 and 14.2 illustrate the consequences of an error in the estimation of the moisture content of the fine aggregate in a mix specified as 320 kg/m³ cement, 6:1 aggregate:cement ratio, 0.55 water/cement ratio, when the assumed moisture content is 5%. If, for example, the moisture content is actually 7% the error in water **addition** is around 12% and the water/cement ratio becomes 0.6. Thus the potential errors are sizeable, and even if they are detected later, water that has been put in cannot be taken out.

Step (d)

Testing of the intermediate product, the fresh concrete, is restricted to subjective assessment based on visual inspection supplemented by occasional slump testing that is carried out not at the plant, but at the delivery site. Barber states[7]: 'The (QSRMC) scheme requires every load to be inspected visually during mixing of the concrete and immediately before delivery. Slump tests are carried out every time a cube is made and the data fed back to the batcher'.

The remark made in Chapter 1 may be repeated here, that the field of view available to the batcher may be very limited or the level of illumination may be inadequate, and attention is drawn again to the other comments already made about subjective assessment.

Step (e)

Testing of the final product, the hardened concrete (as a material), normally depends entirely on the compression testing of concrete cubes that should have been made and cured in a standard way. Quality-control systems based on 28-day strengths, and on correlations between early-age strengths and 28-day strengths, have been devised, and there is no doubt that methods of this type, using for example cusum charts, have contributed substantially to improvements in control of concrete production. The obvious defects are that they can detect only trends, and that the results are available only after the particular batches of concrete have been placed on the job and have set. There is no hope that they can be used to prevent the

Table 14.1 Variability of factors affecting workability and their effects on water content of concrete

Factor that may vary	Possible necessary change in water addition (in litres/m³) if slump is to be kept constant
(a) Coarse aggregate grading	±3
(b) Coarse aggregate shape	±6
(c) Fine aggregate grading	±5
(d) Coarse aggregate moisture content (say variation of 1%)	±12
(e) Fine aggregate moisture content (say variation of 2%)	±15
(f) Absorption of aggregates (normally assumed negligible)	Could be up to ±3
(g) Water demand of cement (change of 30 mm slump not exceptional)	±6
(h) Temperature	±6
(i) Batching accuracy (possible oversanding of 15 kg/m³ say)	+2

Note: Some sources of aggregates may show larger effects than those quoted.

acceptance of defective batches and certainly no possibility of using them for immediate corrective action.

14.3.1 Appraisal

Ready-mixed concrete suppliers are well aware of the problems caused by the variability of materials and they attempt to deal with them. For example, one company has issued a guide from which the information in Table 14.1 has been abstracted. Many ready-mixed plants are equipped with a wattmeter or ammeter connected in the supply line to the mixer and its readings are used in an effort to control workability, and to meet a given slump specification. Since the speed of the mixer is not varied, this is another single-point test subject to all the criticisms that have been given already and, like the slump test, it can classify as identical concretes that are not. Moreover, because it operates at a rate of shear different from that prevailing in the slump test, its results cannot be expected to correlate with slump. The fact that they do not is illustrated by results reported by Wallevik[11] and shown in Figure 14.3; here, the correlation coefficient is only 0.3 for nine degrees of freedom which means that there is no evidence of any correlation.

Exactly the same strictures apply to the slump meter that has been fitted to some ready-mixed concrete trucks. This is a pressure gauge fitted in the hydraulic line of the drum drive and measuring indirectly

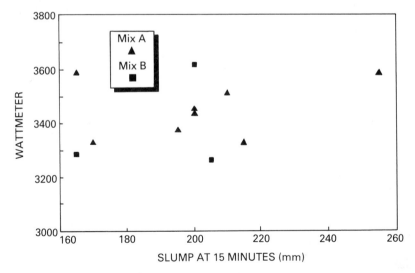

Figure 14.3 Lack of relationship between wattmeter reading and slump value. (*Wallevik*)

the torque required to rotate the drum. The scale of the gauge is marked in terms of slump. This apparatus was developed as a result of work by Harrison[12] and was subsequently patented[13].

Of course, provided only water content changes, and nothing else, these two methods, like other single-point methods, will permit some degree of control but the results will be misleading, and possibly seriously so, as soon as any other factor alters. Some batchermen do appreciate that any correlation they may have between wattmeter reading and slump is different for different mix specifications; in other words, the practical batcherman has recognized that a single-point test is inadequate except for variation of water content only.*

It seems fair to conclude that objective testing of the fresh concrete is so deficient as to be practically non-existent and subjective assessment does not and cannot substitute for it. In the other four of the five steps in establishing a control system, most of what could reasonably be done is already being done and improvements could only be marginal, at least in the case of good plants. The weak link is the inadequacy of testing of the fresh concrete; it is here that attention could be given and could lead to worthwhile improvements in control.

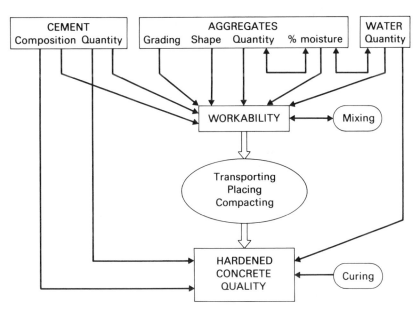

Figure 14.4 Factors affecting properties of concrete.

---

*See Appendix.

14.4 POSSIBILITIES IN TESTING FOR CONTROL

It is worth recalling that the object in controlling concrete production is to produce a material that in its **hardened** state fulfils all that is required of it, and that the properties in the fresh state are of subsidiary, though still very great, importance. Therefore it is not sufficient to control workability alone without consideration of the possible effects on the properties of the hardened concrete.

The important relationships are indicated in Figure 14.4[14] which shows the main factors that, for a given mix specification, can vary and influence the properties of the hardened concrete. Of the seven listed, only three, cement quality, cement quantity, and water quantity, have a major direct effect. The others, which relate to the aggregates, affect the hardened properties principally because they affect workability of the fresh concrete. Changes in them either cause changes in workability so that perhaps the concrete cannot be placed and compacted adequately and there is a consequent loss of strength, or, so that such a consequence can be avoided they necessitate changes in mix proportions. Moreover, the three factors that do have a direct effect on hardened properties also affect workability. Workability also affects, and is affected by, the mixing process so that with the exception of the independent process of curing, the property of workability is related to all the other items shown in Figure 14.4, as well as to other factors, such as the presence of admixtures or replacements, which have been omitted from the figure in the interests of simplicity.

Although both suppliers and purchasers of fresh concrete are well aware of the information summarized in Figure 14.4, it is not uncommon for it to be ignored in practice and for the tacit assumption to be made that water content is the only variable of importance because the others can be taken as being sufficiently constant. If that were true, there would be, for a given mix specification, a simple relationship between results of a single-point workability measurement, such as slump, and the 28-day cube-strength results. There is no published suggestion that the slump test should be used in this way and in fact Shilstone[15] claims to show that, for a very large number of site results, there was no correlation between slump and cube strength. Although Shilstone's statistical treatment is unsatisfactory, his general conclusion is right in essence.

Nevertheless, practical decisions are not infrequently made on the basis that if the slump is right the mix must be right generally, and not only in terms of its workability. Conversely, it may be assumed that if the slump is high the concrete must be defective. An example of this has already been given, of a job where a concrete with a slump considerably higher than specified was delivered and cast into a column, and when this was subsequently discovered by the Resident Engineer he insisted that the column concerned be demolished although non-destructive testing indicated that the suspect concrete was as strong as the rest.

The question that arises from Figure 14.4 is to ask whether it is possible to test workability in such a way that the results can indicate, if workability differs from that specified, what is the cause of the departure. If that objective cannot be fully attained, it would be very useful if batches could be separated into those for which variation of water content is the only factor causing variability, and those for which some other factor or factors has influence. This cannot be done with any single-point test; if slump differs from that specified there is no way of knowing why. The potential of the two-point test will now be examined.

14.5 CONTROL POTENTIAL OF THE TWO-POINT TEST

The potential of the two-point test in matters of quality control lies in the fact that the various factors that can affect workability, such as the properties and proportions of mix components, affect g and h, of course, but do so in different ways according to what is the causative agent. For example, an increase in water content decreases both g and h, whereas an increase in plasticizer content decreases g with comparatively little effect on h. These and other effects have been discussed in more detail in earlier chapters but may be summarized for practical purposes in making site decisions as shown in Table 14.2.

The information in this table is deliberately somewhat simplified but it provides a satisfactory basis for practice. It will be noticed that the first three factors have effects which, at least in direction (i.e. increase or decrease), are the same for all mixes, whereas the others

Table 14.2 Factors affecting workability

Cause Increase in	Result Change in	
	g	h
Water	Decrease	Decrease
Plasticizer	Decrease	None
Air entrainment	None	Decrease
Fines	Depends on particular mix	
Cement	Depends on particular mix	

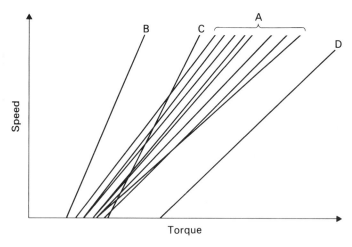

Figure 14.5 Site results. Example (1).

have effects whose size **and** direction depend on the mix under consideration. For example, as was shown in Chapter 9, an increase in fines may increase both g and h, decrease both, or increase one and decrease the other, depending on the mix that is the starting point. A consequence is that an attempt to introduce the two-point test as a quality control method can be looked at in two stages, that is, respectively, without and with preliminary investigations on the mix(es) of interest. The fullest information possible will be obtainable only in the latter case, but considerable progress can be made even in the former case. Examples taken from actual site investigations will make this clear.

14.6 QUALITY ASSESSMENT WITHOUT PRELIMINARY INVESTIGATIONS

All the examples given in this section relate to cases where the two-point apparatus was installed on site and simply used to collect results on a more or less routine basis.

Example 1

Figure 14.5 shows results obtained on eleven nominally identical batches of a superplasticized flowing concrete that was being used to cast a basement floor, and it can be seen that the concrete supplied was very variable. The lines marked A form a fan-shaped set typical of

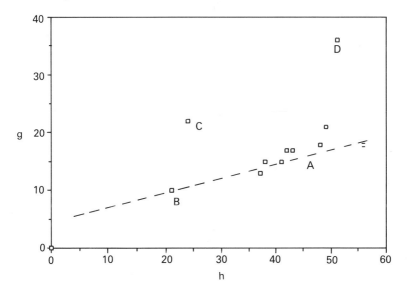

Figure 14.6 Replot of data from Figure 14.5.

the type obtained when only water content is varied. Line B, further to the left, can also be regarded as belonging to this set, and the simplest explanation of the high workability of this batch is that far too much water was added. Line D is quite accurately parallel to one of the lines in the set, i.e. g is higher but h is the same, and the difference in this case can be attributed to a failure to add the correct amount of superplasticizer. Line C crosses several others, and this was thought to be due to a change in the fines content or in the nature of the sand. These observations were communicated to the project manager who had been responsible for obtaining the results. He replied that the batch represented by line D had in fact contained less superplasticizer (by deliberate decision) and he confirmed that during the making of these batches there had been a new delivery of sand, although he could not specifically identify the batch(es) that might have been affected.

An alternative way of considering the results is shown in Figure 14.6 as a plot of g against h. The points for batches A and batch B fall close to a single line with a positive slope, indicating quite clearly that for these batches the only factor contributing to the variability of workability is variability of water content but the points for batches C and D fall well away from the line showing that some factor other than variation of water content is responsible.

This illustrates a simple technique that can be used generally. If

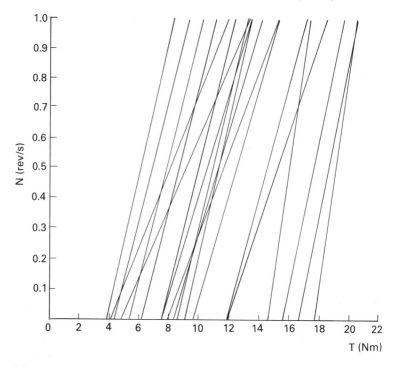

(a) L20 to L39. Morning

Figure 14.7 (caption opposite).

the relation between g and h is found to be a simple line with a positive slope, the factor causing variation is water-content variation only. (Theoretically, complex combinations of other factors could result in such a simple relationship but that occurrence is so improbable that it can be ignored as a practical possibility.) If the relationship between g and h is of any other form, or if there is no apparent relationship, simple variation of water content can immediately be ruled out as the cause of workability variation. When the relationship does exist, individual points distant from the line indicate that, for the batches they represent, some factor other than water content is important.

Example 2

A request was received from a site where considerable trouble had been experienced because of variability of the concrete, for an investigation to be carried out with a view to identifying the causes. The mix as specified by the supplier was $330\,\text{kg/m}^3$ OPC including 50%

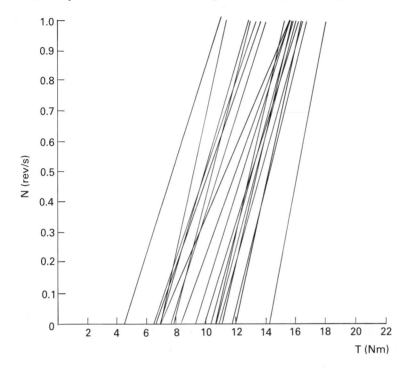

(b) L40 to L60. Afternoon

Figure 14.7 Site results obtained in one day. Example (2).

ggbs, aggregates 20 mm crushed limestone and quartzite sand with A:C 5.5:1 and 43% fines, and water to total cementitious ratio 0.5. The mix also contained a lignosulphonate plasticizer.

Preliminary experiments, conducted mainly to instruct a previously inexperienced operator in the use of the apparatus, gave flow curves that tended to be parallel to each other showing that variability was largely due to variation in the effects of the plasticizer. The supplier was asked to check on his measurement of plasticizer and subsequently 54 results were obtained in one day when all batches delivered to the site were tested. The results are shown in Figure 14.7 and the following deductions can be made.

(a) Batch-to-batch variability is very high.

(b) During one day variability decreased and the variability of batches delivered in the afternoon was statistically significantly less than that of batches delivered in the morning. This suggests that either there was a change in batcherman or that the batcher-

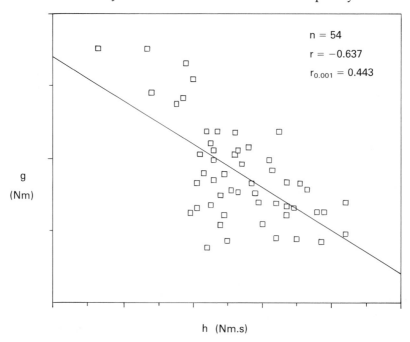

n = 54

r = −0.637

$r_{0.001} = 0.443$

g
(Nm)

h (Nm.s)

Figure 14.8 Results from Figure 14.7 plotted as g against h.

man was gradually learning to deal with this particular mix specification.

(c)   Changes in workability showed no relation to the age of the concrete at the time of test although that age varied from about 15 min to 75 min.

(d)   There was no discontinuous change in workability levels that could be associated with some discontinuous change at the plant, such as delivery of a new batch of aggregates.

(e)   A plot of g against h (Figure 14.8) showed that there was a very highly significant correlation between the two but that correlation was **negative**; this relationship immediately rules out variation of water content as the cause of the variation in workability. The tendency again for the flow curves to show parallelism suggested that the culprit was variation in plasticizer. Because the supplier had been asked specifically to check on the measurement of quantity it was thought the cause was likely to be some other factor such as variation in the time or method of addition. In the original report[16] on this job it was stated that the negative correlation between g and h could be explained if the batcherman was making some error in dealing with the plasticizer and then attempting to correct, on the basis, of subjective judgment or wattmeter reading, by adjusting the water content to compensate. It is now realized that this statement is quite incorrect; such a practice would result in a **positive** correlation, so some other explanation must be sought. This is considered again in the next example.

Example 3

On another site where, again, a lignosulphonate plasticizer was being used, variability was again thought to be due to the plasticizer because the flow curves obtained tended to parallelism. The discharge from the dispenser was therefore checked and, although the quantity was 5% less than indicated by the apparatus it was reproducible. Care was also taken to add the plasticizer in the same way consistently from batch to batch and then it was found that, although the variability of the concretes produced was not great, there was again a statistically significant negative correlation between g and h. As before, this immediately rules out variation in water content and indicates variation in plasticizer as the culprit. But this time, it is known that the amount added was satisfactorily controlled and so was the time and method of addition, and it is also realised that the previously proposed explanation in terms of compensation by altering water content is incorrect. The fact that statistically significant negative correlations have been obtained on several occasions cannot be ignored and it demands some explanation.

Because the correlation is negative, the explanation must be in terms of some factor whose variability causes changes in g and h in opposite directions. That factor might be related to the fact that a lignosulphonate can act as a plasticizer (affecting g preferentially) and also as an air-entraining agent (affecting h preferentially). It is interesting to note that, since this suggestion was first made, Penttala, in the work mentioned in Chapter 10, found that delayed addition of a superplasticizer, caused an increase in workability and a decrease in air content. Although he used melamine formaldehyde and naphthalene formaldehyde sulphonates, his results do lend some support to the suggestion. If there is any substance in it, the effect might be expected to depend on the nature and duration of the mixing process.

On one afternoon mixing time was recorded for 13 batches and was found to have a mean of $3\frac{1}{2}$ min with a range from 1 to 5 min. Workability measurements were made on five of these batches with the results shown in Table 14.3. The correlation coefficients obtained in this case are rather low but again, and this time for only five results,

Table 14.3 Mixing time and g and h for plasticized mix

Mix time (s)	Slump (mm)	r	g (Nm)	h (Nms)
66	70	0.943	8.08	0.72
223	110	0.989	5.25	1.96
300	110	0.972	5.17	1.83
66	70	0.970	6.99	1.64
210	120	0.962	4.16	2.44

there is a statistically significant correlation between g and h with $r = -0.94$. Moreover, as shown in Figure 10.7 of Chapter 10, there appears to be a tendency for g to decrease with mixing time and for h to increase. These two relationships are not statistically significant, chiefly because of the very small number of results available, but there is nothing to be lost by guessing from them that it would be worthwhile to standardize mixing time. This was therefore done and seems to have been justified by the fact that the ranges of g and h were reduced by 34% and 15% respectively.

Example 4

On one piling contract routine measurements were being carried out when several consecutive batches showed a sudden increase in h to about double its normal value, while g remained substantially unaltered. This led to an examination which showed that, in error, the suspect batches contained crushed limestone coarse aggregate instead of the gravel that had been specified. As pointed out earlier, the resulting concrete would be unsuitable for piling because it does not flow satisfactorily.

Example 5

The results shown in Figure 14.9 and 14.10 were obtained with the LM (planetary) form of the apparatus on two separate days and the figures given against the lines are the final figures of the batch numbers: that is, they show the order of delivery of the batches. Although there are only about half a dozen results from each day the following conclusions can be drawn with confidence.
(a)   The patterns of results from the two days are markedly different and sufficiently so as to indicate some definite change in practice

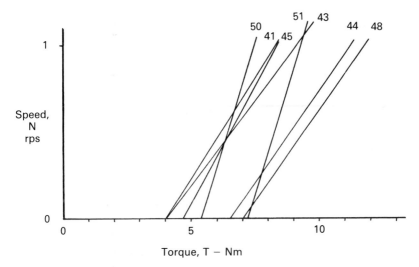

Figure 14.9 Site results. Example (5) Day 1.

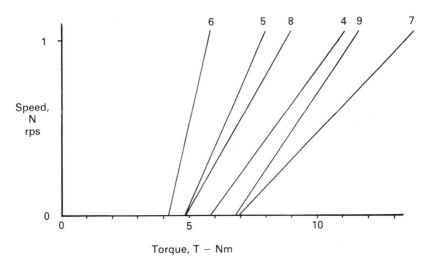

Figure 14.10 Site results. Example (5) Day 2.

in the making of batches, such as might arise from the employ-
ment of a different batcherman.

(b)   Results from the second day form the fan-shaped pattern typical
of the case when only water content is changing and this is confirmed by the fact that the correlation coefficient between g and h

is 0.90 which, even for this small number of results is statistically significant.

(c)   The first five of the seven results from the first day are reasonably parallel to each other and so are the last two. This indicates the cause of variability as being plasticizer variability.

(d)   The last two lines, while being parallel to each other, cross the set of the other five, which, as already said, are also parallel to each other. This sudden change, between batches 48 and 50, suggests that some sudden change took place at the plant. That change could have been, for example, delivery of a new batch of aggregate.

All the examples above show what can be achieved by using the two-point test without any special preliminary experiments but with the exercise of common sense and some knowledge of what else is happening on the site or at the plant. If it is possible to carry out investigations before the job starts, the possibilities become wider.

14.7 QUALITY ASSESSMENT WITH PRELIMINARY INVESTIGATION

Example 6

It has already been shown that the effects of changes in factors such as fines content and cement content vary for different mixes and no generalizations can be made, so if it is desired to exercise some control on these factors it is necessary to obtain information specific to the mix under consideration, preferably before the start of the job. So far, very few opportunities have been offered to carry out trials of this nature so only one example can be given.

Table 14.4 Laboratory trial mixes

Mix No.	Free W/C	Slump (mm)	Fines (%)	Cement (kg/m³)	g (Nm)	h (Nms)	Comment
ST1	0.51	175	40	400	2.80	2.27	Specified
ST2	0.49	165	40	400	2.94	3.35	Less water
ST3	0.56	190	40	400	2.11	1.38	More water
ST4	0.51	203	38	400	1.90	2.71	Less fines
ST5	0.51	159	45	400	3.11	2.42	More fines
ST6	0.54*	209	41	380	1.99	2.91	Less cement
ST7	0.49*	110	41	420	4.41	4.59	More cement

*Same free water content as ST1

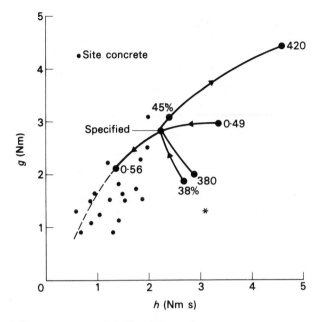

Figure 14.11 Site concrete variability. (*Bloomer*)

Diaphragm walls were being constructed by displacing bentonite in an excavated trench with a mix specified to have a cement content (SRPC) of $400\,kg/m^3$ with an aggregate:cement ratio of 4.5:1 and a W/C ratio of 0.51. The aggregate was a gravel combined with a washed sand at 41% fines and the mix contained a plasticizer. Before the site was visited, trial laboratory tests were carried out on the specified mix and on mixes based on the specified mix, using the same materials as were to be used on site, to get some idea of the effects of possible sources of the expected variability of site results. The mixes and results were as shown in Table 14.4.

During a subsequent visit to the site, tests were carried out on twenty deliveries of concrete; the results are plotted in Figure 14.11, which is due to Bloomer[17]. It can be seen that nearly all the site mixes are more workable than the specified mix as shown by the fact that both g and h are less than they should be. Some of them are very much more workable than specified as was confirmed by the four slump tests carried out by site personnel. In most cases, for which the points lie close to the line showing the effect of change of water content, or to that line extrapolated, the higher workability is simply due to high water content. In the case of the batch marked by an asterisk, the change in workability is probably due to a reduction in cement content or percentage fines and, while it is not possible without further evidence to say which, it would be easy to decide by means of a simple sieve test, in a matter of minutes. For several of the batches some distance to the right of the dotted line, the change in workability is probably caused by a combination of a decrease in fines or cement together with an increase in water content.

It is perhaps worth pointing out also that there are several other mixes on the plot that have approximately the same value of g, and therefore the same slump, as the asterisked batch but the workability of the latter is lower because of the higher value of h. Compared with the specified mix, the value of g for the asterisked mix is lower but the value of h is higher so, in the absence of further information of the type discussed in Chapter 13, it is not possible to say whether on the actual job this mix would be more workable or less workable than the specified mix.

The principles illustrated in this example could obviously be extended, for example by investigating the effect of changes in fines content at more than the one cement content, and a diagram with more information could easily be constructed. The amount of work required is quite small and is easily justified by the potential benefits. Kay's work, described in Chapter 13, is an excellent example.

14.8 PREDICTION OF STRENGTH

It has already been pointed out that, in general, there is no direct relationship between the workability of the fresh concrete and the strength of the hardened concrete, and in this connection, Shilstone's results were mentioned. The reason is that strength is determined primarily by the water/cement ratio, provided the concrete is properly

Table 14.5 28-day cube strengths and workability (Case 1)

g (Nm)	h (Nms)	Strength (N/mm²) (mean of two)
5.06	2.03	49.2
3.17	1.71	46.9
4.10	2.70	55.6
3.16	2.70	47.4
2.89	3.33	53.2
3.39	2.55	49.9
4.41	2.81	50.1

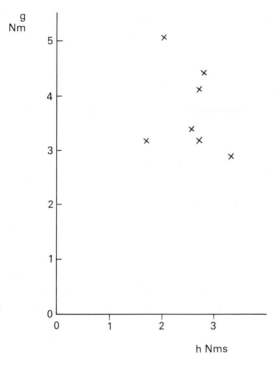

Figure 14.12 Plot of g against h. (Case 1.)

compacted, whereas workability is affected by many other factors as well. However, if it could be shown, for a series of nominally identical batches, that the only (or at least the predominant) factor causing variation of workability is variation of water content, then a simple relationship between workability and strength might be expected. Whether or not the required condition holds is indicated by whether or not there is a positive correlation between the values of g and h.

So far, there has been little opportunity to obtain, from site, both two-point test results and cube strengths on the same batches, so available data are limited, but two practical cases may be considered.

Results on seven batches for the first case are shown in Table 14.5. Figure 14.12 shows immediately that there is no positive correlation between g and h so variation in water content is **not** the predominant cause of variation in workability and therefore no correlation between strength and any measure of workability is to be expected; in fact none exists.

Results on the six batches of the second case are shown in Table 14.6. Figure 14.13 suggests for these results that there might be a posi-

Table 14.6 28-day cube strengths and workability (Case 2)

g (Nm)	h (Nms)	Strength (N/mm²) (mean of two)
5.92	1.56	47.5
3.66	1.30	43.4
2.59	0.79	40.8
3.68	2.06	45.5
3.56	2.19	44.7
5.90	1.63	49.0

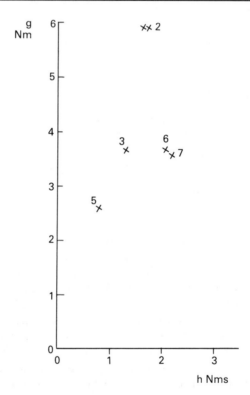

Figure 14.13 Plot of g against h. (Case 2.)

uve correlation between g and h, indicating water content as the main cause of variability, but, for this small number of points, the correlation is not statistically significant. Nevertheless, a plot of 28-day cube strength against g gives the result shown in Figure 14.14 with a correlation coefficient of 0.929 which, even for this small number of results, is significant at the 0.01 level.

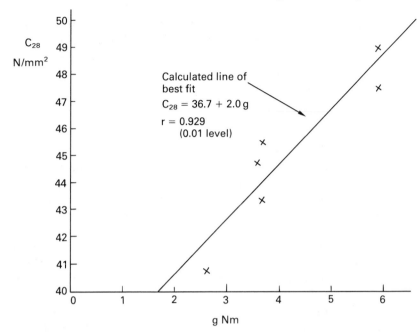

Figure 14.14 Cube strength as a function of workability. (Case 2.)

The standard deviation of the six cube results is 2.94 N/mm² which indicates quite good control, but a figure even more encouraging is obtained by considering the standard deviation about the line shown in Figure 14.14. That line, the best line calculated taking g as the independent variable, has the equation

$$C_{28} = 36.7 + 2.0g \tag{14.1}$$

and the standard deviation about it is only 1.2 N/mm². Thus, most of what variability does exist in this case is accounted for by variation in water content and, if the same degree of control could be maintained, equation 14.1 could be used to predict 28-day strengths from workability measurements on the fresh concrete.

It may also be remarked that, provided the condition exists that water content variation is the predominant cause of variation of workability, strength may also be expected to correlate with the results of a single-point workability test and, in fact, for the results of Case 2 there is a significant negative correlation between strength and slump. Unfortunately, this is of little practical interest because no single-point test can demonstrate that the necessary condition holds to begin with.

Clearly, more work is needed. If the indications of these inadequate results are borne out, it will be possible to predict the strength of hardened concrete to an accuracy of a few N/mm^2 from measurements of workability made before the concrete is placed. The consequential benefits need no emphasizing.

14.9 ESTABLISHING A CONTROL SYSTEM

To establish a control system by testing fresh concrete requires:
(a)   specification of required values of g and h with tolerances;
(b)   testing to determine whether the specification has been met;
(c)   where the specification has not been met, examination of results to indicate reasons for the failure.

The principles underlying all three of these stages have now been discussed and it has been shown how they have been applied in one particular case, that of concrete for pressed slabs as investigated by Kay whose work was discussed in the last chapter. The same methods can be used for any other job if the necessary work is carried out.

Without such work, on site or at the plant, it is not possible to detail a full scheme for any particular job because circumstances vary so widely.

However, it does not follow that nothing can be done in the absence of what some may regard as a major investigation; it is easy to make a start without any preliminary work at all, and to make useful progress simply from the systematic collecting of results from site or plant testing, in addition to carrying out present normal site practice. In any case, it is not to be suggested that here is a ready-made scheme that can be applied without much thought, and that the accumulated knowledge of practising engineers can just be ditched. Introduction of what to most people are new ideas should proceed in stages, and they will be found to be of increasing practical value as they become familiar. Not only do requirements vary from job to job, they also vary at different stages of the job; for example, it is fairly obvious that the purchaser of fresh concrete will be concerned only, or mainly, with stages (a) and (b) listed above, whereas the maker of fresh concrete must necessarily be very concerned with stage (c) if he is to run his plant economically. Bearing all this in mind, a suggested procedure for the introduction of two-point testing on existing jobs might be as follows.
(a)   Carry out **all normal procedures** that would be carried out if two-point testing were not to be introduced. If workability is specified in terms of slump, or any other of the standard tests,

carry out that test as normal. Make all the usual observations and inspections.

(b) Carry out **two-point testing** on as many batches as possible. It is well worthwhile to start with a 'blitz' and test every batch if possible, or every other batch, for a day or two, and then the frequency of testing can be reduced, as experience is gained.

(c) Make sure **batch numbers** are recorded. This may seem obvious but experience has shown that it is often neglected. Record also **age** of concrete at time of test and, if possible, **temperature** of concrete and of surroundings.

(d) Make sure that g and h are obtained for all batches from which **cubes** are taken.

(e) Record an **assessment** on a numerical scale (say 1 to 10) of the suitability of the concrete for the job concerned. This assessment may if possible be made in objective terms by noting, say, pumping pressures (record also the length and nature of the line) but will often have to made in purely subjective terms when it should be done by an experienced engineer and/or by the operative actually working the concrete.

(f) Make sure that g and h are obtained for difficult batches and particularly for any that are **rejected**.

(g) Arrange to have available records of delivery of **materials** such as aggregates, cement and replacements.

14.9.1 Treatment of results and indications

(a) Correlation coefficients should be of the order of 0.99. If they are there is no need to plot the experimental points. For coefficients appreciably less than 0.99 plot the points to see what is causing the relatively poor fit to a straight line. If the highest or lowest point is off the line, reject it and recalculate r, g and h for the remaining points. If other points, or more than one point, are off the line the determination must be accepted as a poor one with a correspondingly high experimental error on g and h.

(b) Plot a control chart for g and h separately; i.e. plot g against batch number and similarly for h. These charts will show the extent of variation of g and h, whether the value of either suddenly increases or decreases, and whether variability increases or decreases. If there is a change of a sudden or systematic nature, see whether it can be associated with some other sudden change such as delivery of a new batch of aggregate. In fact deliveries could be marked on the control charts. If some association seems to be indicated examine the material(s) concerned. For example, if there has been a new delivery of coarse aggregate at about the time that a change occurred, has the new batch a greater or lesser proportion of angular particles than the previous one?

(c) For a suitable period (say each day or half day) use the values of g and h to plot the calculated flow curves (there is no need to plot the experimental points) and examine the pattern.

    (i) Is the pattern for one period different in appearance from that for another period? If so, some systematic change is indicated such as a change at the plant, or a different batcherman.

    (ii) Is the overall spread of the pattern greater or less in one period than in another? If so, in the first case somebody is getting careless or in the second is gradually improving as he learns to deal with the mix concerned.

    (iii) Is the pattern a fan-shaped set? If so, the main cause of variability of the concrete is variation in water content.

    (iv) Do many of the lines tend to be parallel to each other? If so, variation in plasticizer is likely to be a main cause of variability.

    (v) Do the lines up to a certain batch number form a particular pattern and the lines after that batch number form a similar pattern but one that crosses the first? If so, at that batch number some sudden change has occurred such as delivery of new material.

(d) Plot the results as a graph of g against h, and examine the pattern.

    (i) Is there a positive correlation between g and h and is it statistically significant? If so, the cause of variability of the concrete is variation in water content and the points at low values of (g,h) represent concretes that may subsequently fail on cube results.

    (ii) Do most of the points show a positive correlation between g and h and is that correlation statistically significant if a few of the experimental points are omitted? If so, the points in the correlation are as under (i) above; the others represent concretes in which something other than, or in addition to, water content has changed.

    (iii) Is there a negative correlation between g and h? If so, water content can definitely be ruled out as the cause of variability of the concrete. The underlying cause of a negative correlation is not yet fully elucidated but may be associated with changes that affect the effectiveness of a plasticizer. Check the plasticizer dispenser for consistency of delivery; check time of addition of plasticizer; check whether total mixing time is very variable; and also look at temperature records.

(iv)   Is there no correlation between g and h? If so, nothing further can be deduced from this plot.

(e)   On the g v. h plot, enter against each point the suitability number derived from the practical assessment and examine the pattern.

    (i)   Do the batches designated '5' (i.e. the most suitable) tend to fall into a fairly clear zone that can be delineated, with higher and lower numbers tending to fall on one side and the other? If so, the delineated zone gives the combinations of values of g and h that are best for the particular job.

    (ii)   Is the delineated zone a straight line or fairly narrow band? If so, it is possible to calculate the average effective shear rate characteristic of the job, in terms of a speed in the test apparatus. Some of the future tests could then be carried out at that one speed only thus saving time, but at a cost in terms of experimental error.

(f)   From the g v. h plot extract those results that fall on the line that shows water variation only and for the corresponding batches plot cube strength against h. Is there a statistically significant correlation? If so, calculate the best line and use it for predicting cube strengths of future mixes that fall on the particular line showing water variation only.

The system described above can be established without any preliminary work; it requires only the systematic collection of data on the job. Once established it can provide valuable information to assist in decisions on rejection or acceptance of concrete for a particular job and, to a limited extent, can indicate what are the underlying causes of variability. Even this represents a considerable improvement on present practice in quality control and the cost of carrying out an extended trial on the lines suggested is small in comparison with the potential benefits.

At a later stage it might then be felt to be worthwhile to carry out the supplementary work needed for the preparation of proper control charts of the type that will undoubtedly be needed by the maker of the concrete if he is to satisfy any user who has established his own control system. The maker needs to know more about the causes of variability than does the user, who, when the chips are down, is interested only in pass/fail criteria. The additional knowledge can be acquired by preparing charts of the type shown in example 6 or as plotted from Kay's work described in the last chapter. Again, details will depend on the particular job but the principle is to consider what are the most likely factors to vary in practice and then to investigate their effects, singly and in combination. In a complicated case the charts would be prepared in the form of transparent overlays and, of course, the information would eventually be stored in readily accessible form in a computer.

However, undoubtedly, the first step is to apply and experiment with the simple system outlined above. Perhaps it needs to be re-stated that the development of a control system cannot in principle be carried any further by laboratory work, but laboratory work has provided the evidence to show that it is worth trying seriously and that there is promise of considerable financial benefit for those who are prepared to make the relatively small effort necessary.

14.10 REFERENCES

1. Tipler, T.J. (1986) The quest for quality, 1935–1985, Paper presented at seminars on Development of Concrete, London, Edinburgh and Cardiff, Reprint 2/86, Cement & Concrete Association.
2. Quality Scheme for Ready Mixed Concrete *Manual of Quality Systems for Concrete*, The Quality Scheme for Ready Mixed Concrete, Walton-on-Thames May 1984, 23pp. and later edition 1989.
3. Newman, K. (1986) Common quality in concrete construction, *Concrete International*, 37–49.
4. Barber, P. (1990) Specification of concrete the easy way, *Concrete*, **24**(3), 34.
5. Harrison, T. (1989) Introducing designated concrete mixes, *Concrete Quarterly*, Summer 1989, 4–5.
6. Tattersall, G.H. (1989) Problems in the quality assurance of ready mixed concrete, *Concrete*, **23**(11), 22–3.
7. Barber, P. (1989) Quality assurance – a view from the inside, *Concrete*, **23**(11), 24–5 (Reply to Ref. 6).
8. Tattersall, G.H. (1990) Quality assurance of ready mixed concrete, *Concrete*, **24**(2), 5 (letter commenting on Ref. 7)
9. Anthony, P.L. (1975) Moisture measurement in sands. Project Report on Advanced Concrete Technology Course, Cement & Concrete Association.
10. Bell, G. (1981) Getting up-to-date with fresh concrete, *Concrete*, **15**(3), 30–2.
11. Wallevik, O. Private communication to GH Tattersall, 1990.
12. Harrison, O.J. (1964) An investigation into the relationship between concrete workability and the pressures in the hydraulic systems of truck mixers, Technical Report No. 27, Feltham, Middx., Ready Mixed Concrete (UK) Ltd.
13. Ready Mixed Concrete Ltd. Improvements in or relating to apparatus for gauging the consistency of wet concrete. British Patent 1223558.
14. Tattersall, G.H. (1982) Control of high workability concrete, *Concrete*, **16**(3), 21–4.
15. Shilstone, J.M. (1988) Interpreting the slump test, *Concrete International*, 68–70.
16. Tattersall, G.H. (1986) An investigation into site control of concrete by workability measurement. Report BS 84, Department of Building Science, University of Sheffield, 17pp.
17. Bloomer, S.J. (1979) Further development of the two-point test for the measurement of the workability of concrete, PhD Thesis, University of Sheffield.

15 Epilogue

I stated in the Preface that my main purpose in this book was to discuss and elucidate the implications for industry of the fact that the flow properties of fresh concrete are explicable in terms of the Bingham model, and I hope that I have not fallen too far short of achieving that satisfactorily. Many of the arguments will be new to many in the industry and naturally there is reluctance to relegate to second place tests that have been in use for a long time, particularly one such as the slump test that is familiar to concrete technologists in all countries. Nevertheless, the defects of those tests are also well known, as are the practical problems that are a direct product of those defects.

Perhaps the most difficult feature of accepting the Bingham model as a basis for working is the conceptual one of learning to look at workability as needing two constants to describe it instead of one. Of course some effort is needed but there are considerable advantages to be gained. It makes possible explanations of the behaviour of concrete in practice, in handling and transporting processes, and also of phenomena that were previously unexplained. These include the facts that results from the various standard tests do not correlate, that two concretes whose workabilities were thought to be the same behave differently on the job, that values for optimum fines contents depend on the test used to find them, and worst of all, that major errors in batching can remain undetected until too late.

In addition, accepting the Bingham model leads to the development of a test whose advantages over the standard single-point tests are as follows:
(a) It readily differentiates between concretes that might otherwise be wrongly judged as identical in workability.
(b) It can cover almost the whole range of workabilities from low to very high on a single scale of measurement.
(c) An estimated experimental error can be assigned to the results of every separate test.
(d) It is less operator-sensitive.
(e) It provides information about the possible causes of a deviation

from the specified workability and therefore information about factors that may affect the properties of the hardened concrete.

(f)   It can provide information about tendency of a mix to bleed or segregate.

The particular form of apparatus that has been described was developed with working conditions in mind, so it is robust and does not need any laboratory facilities. All the site results that have been given were obtained while using the equipment in the open or in an ordinary hut.

Initial development of the method in terms of a separate piece of apparatus was by deliberate decision but, clearly, it is desirable to incorporate it in the mixing equipment so that results may be obtained, and any necessary corrective action taken, before the concrete is discharged. There are several possibilities and some preliminary work has already been done. In the long run, the method could be part of a system such that results are obtained automatically and fed into a computer controlling the mixing plant so that corrective action could also be automatic. In the meantime, considerable progress can be made on the lines discussed; costing should include consideration of the benefits to be obtained in the avoidance of mistakes and disputes, and in the possibility of reducing the working standard deviation of a plant.

In the text I have presented arguments that I regard as convincing but I have tried to do so in an impersonal way with appropriate back-up of technical information. Perhaps I may be permitted to finish with a comment that is a little less detached. In a recent documentary the construction of a skyscraper costing several hundreds of millions of pounds was shown, and in one shot a technician was carrying out a test with which he claimed he was able to control the water content of the concrete for the bases that were to support the whole weight of the steel-framed building of nearly 50 storeys. To an unbiased observer he seemed to be doing this by making sand pies, like a child on a beach, and watching their movement. Incongruity seems to be an inadequate word to describe this situation.

Glossary

The explanations given here are written with particular reference to the way terms have been used in this book, and are intended only as a help to readers who are not familiar with rheology or statistics. They are not rigorous definitions and should not be regarded as such, or used out of context. Formal definitions of rheological terms may be found in BS 5168:1975 *Glossary of rheological terms*.

RHEOLOGICAL TERMS

In some of these explanations reference will be made to measurements made in the two-point workability apparatus in which torque T is measured as a function of impeller speed N.

Apparent viscosity

Shear stress divided by rate of shear. A measure of apparent viscosity is obtained as $k = T/N$ because T is a measure of shear stress and N is a measure of shear rate. For a *Newtonian* T is proportional to N so k is a constant and is in fact a measure of the **viscosity** itself. For a **Bingham** $T = g + hN$ so $k = g/N + h$ and is therefore different at different values of N, or at different shear rates, as it is for any material other than a Newtonian.

Bingham model

Describes a material that does not flow at all until a particular stress, the **yield stress** or **yield value**, is exceeded, then the stress required for flow is equal to the yield stress plus another term that is proportional to the the shear rate. $T = g + hN$ and g is a measure of the yield value. The parameter h is the constant of proportionality that caters for the effect of shear rate and it is a measure of the **plastic viscosity**.

Consistency

A general term, not subject to simple quantification, to describe the resistance of a material to flow or change of shape. This term has often been wrongly used in concrete technology.

Flow curve

A curve relating stress to rate of shear. In the case of the two-point apparatus it is the curve relating torque to speed of impeller.

Fluidity

The reciprocal of **viscosity**.

Mobility

The reciprocal of **plastic viscosity**.

Newtonian fluid

A material for which the shear stress is always proportional to the shear rate and the constant of proportionality is called the viscosity. In other words, shear stress divided by shear rate is a constant and that constant is the viscosity which is usually denoted by the Greek letter η (pronounced eta).

For a Newtonian in the two-point apparatus T/N is a constant which is a measure of (is proportional to) the viscosity.

Plastic viscosity

One of the parameters of the **Bingham model**. Shear stress = yield value plus plastic viscosity times shear rate. In the two-point apparatus $T = g + hN$ so g is a measure of yield value and h is a measure of plastic viscosity.

Plug flow

Refers to flow in a pipe of a material that possesses a yield value. Because of the existence of the yield value, flow starts near the inner surface of the pipe and the material flows forward as a solid plug whose radius reduces as the pressure causing flow is increased.

Power-law pseudoplastic

A material for which the shear stress is related to the shear rate by a power law equation. In the two-point apparatus this appears as

$$T = AN^b$$

where A and b are constants.

Single-point test

A test in which only one measurement is made at only one shear rate. Such a test is satisfactory only for a material that is characterized by a single constant i.e. for a Newtonian fluid.

For any material whose properties are described by more than one constant (i.e. including concrete) such a test can never give sufficient information.

Thixotropy

A decrease in apparent viscosity as material is sheared followed by a gradual recovery when the material is allowed to rest. One practical example is a suspension of the clay mineral bentonite in water. The suspension may be such that it will not flow at all under its own weight but if it is shaken vigorously it will then flow easily. If left to stand it will 'set' again. Note that thixotropy is a reversible phenomenon; the term has often been used wrongly in concrete technology to describe an irreversible change.

Turbulence

A condition of flow in which the velocity components show random variation.

Viscosity

A term that can be used qualitatively, rather like consistency, to refer to the property of a material to resist deformation increasingly with increasing rate of deformation. When it is used quantitatively it is defined as the shear stress divided by the shear rate, when this is constant. That is, it applies **only** to a Newtonian material and its use for other materials is incorrect. But see also **Apparent viscosity**.

Yield value; yield stress

The minimum stress to start flow in a material. In the two-point test it is measured by the parameter g.

STATISTICAL TERMS

Analysis of variance

See **variance**.

Coefficient of variation

Standard deviation divided by **mean** expressed as a percentage.

Correlation coefficient

A quantity that indicates how well two quantities are related to each other. It is usually denoted by the letter r and can have any value between zero and unity. A value of $r = 1$ means that if one of the two variables is plotted against the other all the points would fit exactly on a straight line, and since the fit cannot be better than that, r can never exceed $r = 1$. The value of r can be negative (between 0 and -1) and the negative sign simply means that as one of the variables increases the other decreases. The actual value of r must be considered in relation to the number of **degrees of freedom**.

Degrees of freedom

The number of degrees of freedom is an expression of the amount of information available. No general statement can be given here but, for example, in the consideration of one variable only the number is one less than the number of data points available. If ten cube-strength results are being considered, the number of degrees of freedom (d.o.f.) is nine. When the relationship between two variables is being considered, as for example, torque and speed in the two-point test, the number of d.o.f. is two less than the number of points available.

Distribution curve

A curve showing how some particular quantity varies. Results may first be plotted as a histogram, or bar chart, as shown, for example in Figure 13.1 of Chapter 13 for slump results. If the total number of results is large, the width of the bars may be reduced and then the tops of the bars may be joined together to produce a distribution curve.

Equation to a line

The equation to a straight line is usually calculated by the **least squares** method.

Histogram

See **Distribution curve.**

Interaction term

Variables are said to interact when the effect of one depends on the level of another. For example, as shown by the work of Ellis and Wimpenny on ggbs discussed in Chapter 11, the effect of slag replacement level on g depends on the cement content. At $300\,\mathrm{kg/m^3}$ cement, slag replacement level has little or no effect but at higher and lower cement contents there is an effect. Thus there is an interaction between slag replacement level and cement content. In other concretes, the effect on workability of a change in fines content is different at different cement contents; there is an interaction between fines content and cement content.

Least squares line

Consider a straight-line relationship between y as the dependent variable and x as the independent variable. The distance of each experimental point from the line, measured in the vertical direction along the y-direction, is called the deviation on y. The best line, known as the least-squares line, is the one for which the sum of the squares of these deviations on y is a minimum. It can be calculated by quite simple formulae but a cheap calculator will give the values of intercept and slope if the various values of x and y are fed in. Note that if x and y are interchanged the result will be different; that is, the best line for predicting y if x is known is not the same as the best line for predicting x if y is known. (See comments about equations relating slump and flow in Chapter 2, equation 2.1.)

Level of significance

In statistics, results are usually quoted to a level of significance, which indicates the probability that a given result could have occurred by chance. For example, if a result is significant at the 0.001 level, that means it could have happened by chance 1 in 1000 times which is unlikely. The conclusion then would be that a real effect (and not just chance) has been detected. On the other hand, if the level of significance is 0.1, or 1 in 10, that is a fairly high chance that may well have happened so the conclusion would be that no real effect has been established. The level of significance that would be accepted in practice depends on the consequences of the practical decision to be made. If the consequences of a wrong decision would be serious then a highly significant result (say 0.001 level) would be looked for before the decision would be made. The significance of a **correlation coefficient** depends not only on its own value but also on the number of **degrees of freedom**. A value of $r = 0.75$ with 5 degrees of freedom would be significant only at the 0.05 level, but the same value of r with 15 degrees of freedom would be significant at the 0.001 level.

Mean

The sum of all the values of a variable divided by the number of values (commonly known as the average).

Median

The middle value when all the observed values are arranged in ascending order. If the number of values is even, it is the mean of the centre two.

Normal distribution

A particular form of distribution which is symmetrical about the mean value so the mean and the median are one and the same. Also known as a Gaussian distribution. It is the one most commonly met in concrete technology and even if a distribution has some degree of **skewness** it can often be treated without too much error as a normal distribution.

Randomization

A series of experiments whose results are destined for statistical analysis should normally be carried out in a random order, that is, in an order determined by chance. There are several ways of selecting such an order but it could be done just by pulling numbers from a hat. This is done to eliminate possible unknown systematic changes. For example, if Al-Shakhshir had not randomized the order of his experiments

(Chapter 10) and had carried out all those on ordinary Portland cement first and then all those on rapid-hardening cement, he could not have been sure that the differences he found were due to the change in cement because there might have been some other unknown change (e.g. ambient temperature) that had occurred between the two series. By randomizing he eliminated this possibility and any effect due to an unknown variable would be included in his estimate of overall experimental error.

Skewness

A measure of the departure of a distribution from symmetry. Certain types of distribution, such as the Poisson distribution, are quite definitely skew.

Standard deviation

A measure of the width of a distribution, that is, of the variability of the quantity being studied. Usually designated by the Greek letter σ (lower case sigma), sometimes as SD or simply S. It occurs as a parameter in the equation for the curve representing a normal distribution. It is defined as the root mean square deviation and may be calculated as follows: from each value of the variable x (say) subtract the mean value of all the x's to get the deviations, square each of those deviations, sum those squares, take the mean by dividing by the total number of x's, then take the square root. There are simpler formulae and a cheap calculator will produce the result if all the values of x are entered. In a normal distribution about 68% of the values lie within a distance from the mean equal to the standard deviation S, (plus and minus) and about 95% lie within a distance of ±2S of the mean.

Variance

Also a measure of the width of a distribution. It is equal to the square of the standard deviation. It is a useful quantity because variances are additive. For example, if a measured quantity, such as cement content of a batch, is affected by sampling error and testing error, the variance of the final result is equal to the sum of the sampling variance and the testing variance. Analysis of variance refers to methods of splitting a total variance into its component parts provided the experiments have been carried out to a suitable design. For example, an analysis may be carried out to investigate the relative importance in their effects on a property of concrete of such factors as cement content, fines content, replacement level etc (see also *interaction term*).

Appendix: BS 5328:1990

BS 5328:1981 Methods for Specifying Concrete, discussed in the text, has recently been superseded by a new edition published at the end of 1990. This new edition is in four parts:

Part 1 – Guide to specifying concrete;

Part 2 – Methods for specifying concrete mixes;

Part 3 – Specification for the procedures to be used in producing and transporting concrete;

Part 4 – Specification for the procedures to be used in sampling, testing and assessing compliance of concrete.

The main changes in relation to workability are summarized below.

1. *Workability requirements*

At variance with what might be thought to be the over pessimistic opinion expressed in 'Design of normal concrete mixes' (quoted in Chapters 1 and 13) that "it is not considered practical . . . to define the workability required for various types of construction or placing conditions . . .", some guidance is now given in a table, shown here as Table A1. Indication is also given of the methods of compaction to be used.

2. *Control of water content*

Clause 4.7 of Part 3 permits the control of water content by controlling workability provided that the relationship between water content and workability for the materials used is available. This clause also requires that "Batchermen and mixer-drivers shall be made aware of the appropriate responses to variations in concrete consistence of a particular mix caused by normal variations in aggregate moisture content or grading".

(See Chapters 1 and 14 for discussions of subjective assessment and control.)

3. *Temperature*

In the earlier edition a minimum concrete temperature of 5°C was required, and now, with reference to hot weather working, it is also specified that temperature of the concrete at delivery shall not exceed 30°C (Clause 4.9 of Part 3). Variations on these minimum and maximum temperatures may be permitted or specified by the purchaser. The clause also gives other practical requirements for hot and cold weather working.

(See Chapter 8 for discussion of the effect of temperature on workability.)

4. *Segregation*

Clause 4.10 of Part 3, which deals with transport of concrete, lays down that the method used shall "prevent segregation" and also states, "The concrete shall be deposited as close as possible to its final position to avoid rehandling or moving the concrete horizontally by vibration".

(See Chapter 9 for discussion of segregation.)

5. *Specification of workability*

An addition has been made to the information given as Table 13.1 of Chapter 13 in that Clause 3.5 of Part 4 says that the tolerance to be allowed on a specified flow table value shall be ± 50 mm.

Note that, according to Sym, the standard deviation for this test is 25 mm (see Chapter 2), so that for a tolerance of ± 50 mm the value of Sym's Test Capability Index (see Chapter 14) is

$$TCI = (2 \times 50)/25 = 4.$$

Sym says that a value as low as four means that the test is unsuitable for the intended purpose. As pointed out in Chapter 14, a value of TCI = 4 means that 1 in 20 of the batches tested will fail because of variability of the test itself.

Table A1 Workabilities suitable for different uses of *in situ* concrete (from BS 5328: Part 1: 1990)

Use of concrete	Form of compaction	Workability	Nominal slump*
			mm
Pavements placed by power operated machines	Heavy vibration	Very low	See Note 1
Kerb bedding and backing	Tamping		
Floors and hand placed pavements	Poker or beam vibration	Low	50
Strip footings Mass concrete foundations Blinding Normal reinforced concrete in slabs, beams, walls and columns Sliding formwork construction Pumped concrete Vacuum processed concrete Domestic general purpose concrete	Poker or beam vibration and/or tamping	Medium	75
Trench fill *In situ* piling	Self-weight compaction	High	125
Concrete sections containing congested reinforcement	Poker		
Diaphragm walling Self-levelling superplasticized concrete	Self-levelling	Very high	See Note 2

*Cohesive mixes may give adequate placeability at lower values of slump than those given here.

Note 1. In the 'very low' category of workability where strict control is necessary, e.g. pavement quality concrete placed by 'trains', measurement of workability by determination of compacting factor or Vebe time (see BS 1881: Parts 103 and 104) will be more appropriate than slump.

Note 2. In the 'very high' category of workability, measurement and control of workability by determination of flow will be appropriate (see BS 1881: Part 105).

Conversion factors

Length

1 in	= 25.4 mm	1 mm	= 0.0394 in
1 ft	= 0.305 m	1 m	= 3.28 ft
1 yd	= 0.94 m	1 m	= 1.09 yd

Area

1 in^2	$= 645 \text{ mm}^2$	1 mm^2	$= 0.00155 \text{ in}^2$
1 ft^2	$= 0.0929 \text{ m}^2$	1 m^2	$= 10.8 \text{ ft}^2$
1 yd^2	$= 0.836 \text{ m}^2$	1 m^2	$= 1.20 \text{ yd}^2$

Volume

1 in^3	$= 16400 \text{ mm}^3$	1 mm^3	$= 0.0000610 \text{ in}^3$
1 ft^3	$= 0.0283 \text{ m}^3$	1 m^3	$= 35.3 \text{ ft}^3$
1 yd^3	$= 0.765 \text{ m}^3$	1 m^3	$= 1.31 \text{ yd}^3$

Capacity

1 pt (UK)	= 0.568 litre	1 litre	= 1.76 pt (UK)
1 pt (US)	= 0.473 litre	1 litre	= 2.11 pt (US)
1 gal (UK)	= 4.55 litre	1 litre	= 0.220 gal (UK)
1 gal (US)	= 3.79 litre	1 litre	= 0.264 gal (US)

Mass

1 oz	= 28.3 g	1 g	= 0.0353 oz
1 lb	= 0.454 kg	1 kg	= 2.20 lb
1 cwt	= 50.8 kg	1 kg	= 0.0197 cwt
1 ton	= 1.016 tonne	1 tonne	= 0.984 ton
1 bag (US)	= 94 lb = 42.6 kg		

Force

1 lbf	= 0.536 kgf	1 kgf	= 2.20 lbf
1 lbf	= 4.45 N	1 N	= 0.225 lbf

Stress

1000 lbf/in^2	$= 0.703 \text{ kgf/mm}^2$	1 kgf/mm^2	$= 1420 \text{ lbf/in}^2$
1000 lbf/in^2	$= 6.90 \text{ Nmm}^2$	1 N/mm^2	$= 145 \text{ lbf/in}^2$
1 Pa	$= 1 \text{ N/m}^2$	1 MPa	$= 1 \text{ N/mm}^2$

Capacity per unit volume

1 pt (US)/yd^3 = 0.619 litre/m^3 1 litre/m^3 = 1.62 pt (US)/yd^3
1 gal (US)/yd^3 = 4.95 litre/m^3 1 litre/m^3 = 0.202 gal (US)/yd^3

Capacity per mass

1 pt (US)/cwt = 0.466 litre/50 kg   1 litre/50 kg = 2.15 pt (US)/cwt

Mass per unit volume

1 lb/ft^3 = 16.0 kg/m^3 1 kg/m^3 = 0.205 lb/ft^3
1 lb/yd^3 = 0.593 kg/m^3 1 kg/m^3 = 1.69 lb/yd^3
1 bag/yd^3 (US) = 55.7 kg/m^3 100 kg/m^3 = 1.79 bags/yd^3

Temperature

1 deg F   = 0.556 deg C   1 deg C   = 1.80 deg F
°F   = $\frac{9}{5}$°C + 32   °C   = $\frac{5}{9}$ (°F − 32)

Author Index

Subject Index